1964

SCIENCE, FAITH AND SOCIETY

MICHAEL POLANYI

THE UNIVERSITY OF CHICAGO PRESS
CHICAGO AND LONDON

LIBRARY
FLORIDA KEYS COMMUNITY COLLEGE
5901 West Junior College Road
Key West, Florida 33040

ISBN: 0-226-67289-1 (clothbound); 0-226-67290-5 (paperbound)
The University of Chicago Press, Chicago 60637
The University of Chicago Press, Ltd., London

© 1946 by Michael Polanyi. Published 1946 by Geoffrey Cumberlege
Oxford University Press, London
Introduction © 1964 by The University of Chicago
All rights reserved. Published 1964
Printed in the United States of America

93 92 91 90 89 88 12 11 10 9

CONTENTS

APPENDIX

BACKGROUND AND PROSPECT

IN AUGUST 1938 the British Association for the Advancement of Science founded the Division for the Social and International Relations of Science, which was to give social guidance to the progress of science. A movement for the planning of science spread and became predominant among scientists interested in public affairs. A small number of scientists, to which I belonged, strenuously opposed this movement.

In December 1945 the Division called a meeting to discuss planning and asked me to open the proceedings. My address renewed my criticism of planning and upheld the traditional independence of scientific enquiry. I expected a hostile reaction, but, to my surprise, speakers and audience showed themselves in favour of science pursued freely for its own sake. Since then the planning movement has dwindled to insignificance in Britain, but the theoretical problems it has raised are still with us. They are part of the general impact made by the Russian Revolution on the minds of men everywhere.

After the Revolution, scientific research in Soviet Russia was divided into two sections. One was conducted in the light of dialectical materialism under the leadership of the Communist Academy founded in 1926. Membership in the Academy was confined to Party members. Scientists forming the other section worked freely in constant touch with Western scientists. In 1932 a change occurred affecting both sides. The Soviet government repudiated the wild dialectical speculations of the Communist Academy and covered them with ridicule. At the same time, however, the other part of science, hitherto conducted on traditional lines, was bidden to acknowledge the supremacy of dialectical materialism. A declaration to this effect can be found in the editorial opening of the new German-language physics journal of the Soviets, founded in that year; it was inserted at the request of the Party. Russia's most distinguished biologist, N. I. Vavilov, was induced, in the same year, to denounce the theoretical pursuit of genetics practised in the West and to accept instead the view of science planned to serve economic needs, as declared by the Conference on Planning Genetics Selection Research in Leningrad.

At Easter 1935 I visited N. I. Bukharin in Moscow. Though he was heading for his fall and execution three years later, he was still a leading theoretician of the Communist party. He explained to me that the distinction between pure and applied science, made in capitalist countries, was due to the inner conflict of this type of society which deprived scientists of the consciousness of their social functions, thus creating in them the illusion of pure science. Accordingly, Bukharin said, the distinction between pure and applied science was inapplicable in the Soviet Union. This implied no limitation on the freedom of research; scientists would follow their interests freely in the U.S.S.R., but, owing to the internal harmony of socialist society, they would inevitably be led to lines of research which would benefit the current Five Year Plan. The comprehensive planning of all research was to be regarded merely as a conscious confirmation of the pre-existing harmony between scientific and social aims.

In 1935 I could still smile at this dialectical mysterymongering, never suspecting how soon it would show terrible consequences. Vavilov's persecution at the hands of T. D. Lysenko had already begun. It led to his dismissal from office in 1939 and then to his arrest and death in a prison camp around 1943. This campaign wrought havoc among biologists and paralysed whole branches of biology in Soviet Russia from 1939 until well after Stalin's death in 1953. The physical sciences got off more lightly. By the time of this writing, the natural sciences have been almost completely liberated from ideological subservience to Marxism, which continues to be imposed on the study of economics, sociology and the humanities.

I have said that in England the campaign for the planning of science, evoked by the enforcement of Marxist philosophy in the Soviet Union, never became a serious menace. But the mental disturbance caused by it was profound. A distinguished scientist like Lancelot Hogben could write:

> From the landman's point of view the earth remained at rest till it was discovered that pendulum clocks lose time if taken to places near the equator. After the invention of Huyghens, the earth's axial motion was a socially necessary foundation for the colonial export of pendulum clocks.

Many such absurd theories were put forward in Hogben's famous book *Science for the Citizen* (1938), which had a vast circulation. An account of the considerable literature moving on similar lines is given in my book *The Logic of Liberty* (1951).

It was difficult to get a hearing for opposing views. Those who knew about the persecution of biologists in Soviet Russia would not divulge their information. My writings and those of J. R. Baker which, from 1943 on, exposed this persecution were brushed aside as anti-Communist propaganda. The way in which scientific research was organised in Soviet Russia was held up as an example to be followed. Public meetings, attended by distinguished British scientists, gave currency to this appeal.

It was in facing these events that I became aware of the weakness of the position I was defending. When I read that Vavilov's last defence against Lysenko's theories, in 1939, was to evoke the authority of Western scientists, I had to acknowledge that he was appealing to one authority against another: to the authority accepted in the West against the authority accepted in Soviet Russia. The meeting had been called by the editors of the journal *Under the Banner of Marxism*. Their acceptance of Lysenko's authority was based on their philosophy of science. What philosophy of science had we in the West to pit against this? How was its general acceptance among us to be accounted for? Was this acceptance justified? On what grounds?

Marxism has challenged me to answer these questions: the essays republished here were written in reply to them. Like the Marxist theory, my account of the nature and justification of science includes the whole life of thought in society. In my later writings it is extended to a cosmic picture. But the ultimate justification of my scientific convictions lies always in myself. At some point I can only answer, 'For I believe so'. This is why I speak of Science, Faith and Society.

I first analysed the process of knowing, as is usual, in isolation. There are an infinite number of mathematical formulae which will cover any series of numerical observations. Any additional future observations can still be accounted for by an infinite number of formulae. Moreover, no mathematical function connecting instrument readings can ever constitute a scientific theory. Future instrument readings cannot ever be predicted. But this is merely a symptom of a deeper inadequacy, namely, that the explicit content of a theory fails to account for the guidance it affords to future discoveries. To hold a natural law to be true is to believe that its presence will manifest itself in an indeterminate range of yet unknown and perhaps yet unthinkable consequences. It is to regard the law as a real feature of nature which, as such, exists beyond our control.

We meet here with a new definition of reality. Real is that which is expected to reveal itself indeterminately in the future. Hence an explicit statement can bear on reality only by virtue of the tacit coefficient associated with it. This conception of reality and of the tacit knowing of reality underlies all my writings.

If explicit rules can operate only by virtue of a tacit coefficient, the ideal of exactitude has to be abandoned. What power of knowing can take its place? The power which we exercise in the act of perception. The capacity of scientists to perceive the presence of lasting shapes as tokens of reality in nature differs from the capacity of our ordinary perception only by the fact that it can integrate shapes presented to it in terms which the perception of ordinary people cannot readily handle. *Scientific knowing consists in discerning Gestalten that are aspects of reality.* I have here called this 'intuition'; in later writings I have described it as the tacit coefficient of a scientific theory, by which it bears on experience, as a token of reality. Thus it foresees yet indeterminate manifestations of the experience on which it bears.

Every interpretation of nature, whether scientific, non-scientific or anti-scientific, is based on some intuitive conception of the general nature of things. In the magical interpretation of experience we see that some causes which to us are massive and plain (such as a stone's smashing a man's skull) are regarded as incidental or even irrelevant to the event, while certain remote incidents (like the passing overhead of a rare bird) which to us appear to have no conceivable bearing on it are seized upon as its effective causes. Such a general system may resist many facts which to those who do not believe in the system seem to refute it. Any general view of things is highly stable and can be effectively opposed, or rationally upheld, only on grounds that extend over the entire experience of man. The premisses of science on which all scientific teaching and research rest are the beliefs held by scientists on the general nature of things.

The influence of these premisses on the pursuit of discovery is great and indispensable. They indicate to scientists the kind of questions which seem reasonable and interesting to explore, the kind of conceptions and relations that should be upheld as possible, even when some evidence seems to contradict them, or that, on the contrary, should be rejected as unlikely, even though there was evidence which would favour them.

The premisses of science are subject to continuous modifications. In the appendix to these lectures I have described a series of stages through which the premisses of physics have passed since Copernicus. Every established proposition of science enters into the current premisses of science and affects the scientist's decision to accept an observation as a fact or to disregard it as probably unsound. To show this, a long series of such cases is given in the appendix, and many other examples can be found in my later writings. This material refutes the widely held view that scientists necessarily abandon a scientific proposition if a new observation conflicts with it. The material collected in the appendix also refutes the view that the progress of science affects only the interpretation of the facts and leaves the accepted facts unchanged.

All this is accounted for by the view that the advancement of science consists in discerning Gestalten that are aspects of reality. We know that perception selects, shapes and assimilates clues by a process not explicitly controlled by the perceiver. Since the powers of scientific discerning are of the same kind as those of perception, they too operate by selecting, shaping and assimilating clues without focally attending to them. Thus it is ultimately left to the personal judgement of the scientist to decide what conflicting evidence invalidates a proposition, what things coming to his notice must be accepted as facts and what should be concluded from them.

Gestalt psychology and, more recently, transactional psychology have studied the shaping of percepts. This process consists in our selecting from the material presented to us and supplementing it. The result is an interpretation of the material which may be either compelling or to some extent optional.

The criteria of such shaping are qualitative, undefinable and often conflicting. This applies also to the shaping of experience by science. All great discoveries are beautiful, but the quality of beauty varies. The discovery of Neptune was a brilliant confirmation of hitherto accepted views, the discovery of radioactivity a dazzling revolution against them; each was beautiful in its own way. In *Personal Knowledge* I told of discoveries in mathematical physics, guided by pure theoretical beauty. In a recent paper entitled "The Evolution of the Physicist's Picture of Nature" (*Scientific American*, CCVIII [May, 1963]), P. A. M. Dirac emphatically confirms this: '. . . It is more important to have beauty in one's equations than to have them fit experiment'. I shall presently say more about the final arbitrament of such rival claims.

Maurice Merleau-Ponty's *La Phénoménologie de la Perception* (Paris, 1945) reached this country after these lectures were delivered. The book does not deal with the philosophy of science; yet by analysing perceived knowledge on the lines of Husserl, it arrives at views akin to those I have expressed here. A. D. Ritchie, who was for a number of years my colleague in Manchester, independently developed in *Essays in Philosophy* (London, 1948) and in *History and Methods of the Sciences* (Edinburgh, 1958), ideas on the nature of science basically akin to my own. Of later writers whose conclusions overlap my own, I shall cite W. I. Beveridge, J. D. Bronowski, Stephen Toulmin, N. R. Hanson, Konrad Lorenz, Thomas Kuhn, Gerald Holton, Ch. Perelman and A. I. Wittenberg.

The Art of Scientific Discovery (1950), by W. I. Beveridge, brought invaluable sketches drawn from life to illustrate scientific discovery as an art. J. Bronowski, in *Science and Human Values* (1956), has also developed the view that scientific discovery is a creative act akin to creation in the arts. In *The Philosophy of Science* (1953) Stephen Toulmin has shown systematically that the framework of scientific theories contains general suppositions which cannot be put directly to an experimental test of truth or falsity. Such general premisses overlap more specific statements which embody them. N. R. Hanson has observed in *Patterns of Discovery* (1958) that scientific facts are 'theory-laden'. An essay by Konrad Lorenz, 'Gestalt

Perception as Fundamental to Scientific Knowledge',[1] illuminatingly develops the analogy between the perception of Gestalt and the knowledge of science but does not enquire into the ultimate justification of science, which I have approached from this starting point in the present lectures (1946) and in my *Personal Knowledge* (1958). Thomas Kuhn, in *The Structure of Scientific Revolutions* (1962), pointed out that some major discoveries have profoundly affected the outlook of scientists, and he called these discoveries 'paradigmatic'. Gerald Holton in an essay published in *Eranos Jahrbücher*, XXXI (1962), under the title 'Über die Hypothesen welche der Naturwissenschaft zugrunde liegen', demonstrated the 'thematic' dimension of scientific propositions, which is what I described as their part in embodying general premisses of science. In *Vom Denken in Begriffen* (Basel and Stuttgart, 1957), A. I. Wittenberg shows that reason discovers and must acknowledge, in mathematics, an ultimate knowledge, the content of which cannot be fully explicited. This situation forms part of our intellectual existence. Ch. Perelman, in *La Nouvelle Rhetorique, Traité de l'Argumentation* (Paris, 1958), proceeds from the dubitability of all inferences to an enquiry into the convincing power of rhetorical argument, with which he abides. Wittenberg and Perelman both enquire, as I have done, into the role of decision and personal judgement in science and acknowledge their comprehensive powers. They would seem to share my view, that our dependence on these powers is the fundamental problem of epistemology.

Having dealt with the tacit coefficient of explicit scientific knowledge, we must now turn to the tacit process by which scientific knowledge is discovered. What do we know about the process of scientific intuition?

Surprising discoveries are often made on the grounds of observations that have been known for some time. Jeans quotes as examples the work of Copernicus, Galileo, Kepler, Newton, Lavoisier and Dalton, to which I would add Darwin's work, De Broglie's wave theory, Heisenberg's and Schrodinger's quantum-mechanics and Dirac's theory of the electron and

[1] English translation of 'Gestaltwahrnehmung als Quelle Wissenschaftlicher Erkenntnis', *Zeit. f. exp. u. angew. Psychol.*, 1959, No. 6, 118–65, in *General Systems*, Vol. VII (1962), ed. L. von Bertalanffy and A. Rappaport [Ann Arbor, Mich.].

positron. These inferences from known facts had to await the action of exceptional intuitive powers, and they clearly demonstrate the existence of such powers.

But in spite of much beautiful work done by Gestalt psychologists on problem-solving, of striking descriptions of the process of discovery by Poincaré and by Hadamard and of the pioneering enquiries of Polya into the heuristics of mathematics, we still have no clear conception of how discovery comes about. The main difficulty has been pointed out by Plato in the *Meno*. He says that to search for the solution of a problem is an absurdity. For either you know what you are looking for, and then there is no problem; or you do not know what you are looking for, and then you are not looking for anything and cannot expect to find anything. If science is the understanding of interesting shapes in nature, how does this understanding come about? *How can we tell what things not yet understood are capable of being understood?* The answer I gave here to this question was that we must have a foreknowledge sufficient to guide our conjecture with reasonable probability in choosing a good problem and in choosing hunches that might solve the problem. A potential discovery may be thought to attract the mind which will reveal it—inflaming the scientist with creative desire and imparting to him intimations that guide him from clue to clue and from surmise to surmise. The testing hand, the straining eye, the ransacked brain, may all be thought to be labouring under the common spell of a potential discovery striving to emerge into actuality. I feel doubtful today about the role of extra-sensory perception in guiding this actualisation. But my speculations on this possibility illustrate well the depth that I ascribe to this problem.

Admittedly, there are rules which give valuable guidance to scientific discovery, but they are merely *rules of art*. The application of rules must always rely ultimately on acts not determined by rule. Such acts may be fairly obvious, in which case the rule is said to be precise. But to produce an object by following a precise prescription is a process of manufacture and not the creation of a work of art. And likewise, to acquire new knowledge by a prescribed manipulation is to make a survey and not a discovery. The rules of scientific enquiry leave their own application wide open, to be decided by the scientist's judgement. This is his major function. It includes the finding of a good problem, and of the surmises to pursue it, and the recognition of a discovery that solves it. In each such decision the scientist may rely on the support of a rule; but he is then selecting a rule that applies to the case, much as the golfer chooses a suitable club for his next stroke.

Viewed from outside, as I have just described him, the scientist may appear as a mere truth-finding machine steered by intuitive sensibility. But this view overlooks the curious fact that from beginning to end he is himself the ultimate judge in deciding on each consecutive step of his enquiry. He has to arbitrate all the time between his own passionate intuition and his own critical restraint of it. The reach of these ultimate decisions is wide: the great scientific controversies show the range of basic questions which may remain in doubt after all sides of an issue have been examined. The scientist must decide such issues, left open by opposing arguments, in the light of his own scientific conscience. My book *Personal Knowledge* (1958) attempts to buttress this final commitment against the charge of subjectivity.

Since an art cannot be precisely defined, it can be transmitted only by examples of the practice which embodies it. He who would learn from a master by watching him must trust his example. He must recognise as authoritative the art which he wishes to learn and those of whom he would learn it. Unless he presumes that the substance and method of science are fundamentally sound, he will never develop a sense of scientific value and acquire the skill of scientific enquiry. This is the way of acquiring knowledge, which the Christian Church Fathers described as *fides quaerens intellectum*, 'to believe in order to know'.

To learn an art by the example of its practice is to accept an artistic tradition and to become a representative of it. Novices to the scientific profession are trained to share the ground on which their masters stand and to claim this ground for establishing their independence on it. The imitation of their masters teaches them to insist on their own originality, which may oppose part of the current teachings of science. It is inherent in the nature of scientific authority that in transmitting itself to a new generation it should invite opposition to itself and assimilate this opposition in a reinterpretation of the scientific tradition.

Enforcement of discipline, combined with inducement to dissent, is also exercised by science in controlling the resources of scientific research and the organs of scientific publicity. A contribution to science is accepted only if, in the light of scientific beliefs about the nature of things, it appears sufficiently plausible. Only thus can contributions of cranks, frauds and bunglers be prevented from flooding scientific publications and corrupting scientific institutions. At the same time, scientific authority ascribes the highest merit to originality, which may dissent to some extent from the established teachings of science. This internal tension and its dangers are inevitable.

The authority of science resides in scientific opinion. Science exists as a body of wide-ranging authoritative knowledge only so long as the consensus of scientists continues. It lives and grows only so long as this consensus can resolve the perpetual tension between discipline and originality. Every succeeding generation is sovereign in reinterpreting the tradition of science. With it rests the fatal responsibility of the self-renewal of scientific convictions and methods. To speak of science and its continued progress is to profess faith in its fundamental principles and in the integrity of scientists in applying and amending these principles.

Each scientist is confronted with the criticism of his neighbours, who in their turn are criticised by their own neighbours. Thus the chain of mutual appreciation spreads throughout the body of science, from mathematics to medicine, and maintains the same fundamental beliefs and standards of scientific interest everywhere. Rooted in the same tradition as his colleagues, each scientist independently plays his part in maintaining this tradition over an immense area of scientific enquiry of which he knows next to nothing.

There are differences in rank between scientists, but these are of secondary importance: everyone's position is sovereign. The Republic of Science realises the ideal of Rousseau, of a community in which each is an equal partner in a General Will. But this identification makes the General Will appear in a new light. It is seen to differ from any other will by the fact that it cannot alter its own purpose. It is shared by the whole community because each member of it shares in a joint task. This community would instantly dissolve if this task came to an end and the members of the community had to decide on doing something else.

We can generalise this to other modes of discovery in literature, in the arts, in politics. All these can advance only fragmentarily by the efforts of individuals within a community organised essentially on the lines of scientific life. The community must guarantee the independence of its active members in the service of values jointly upheld and mutually enforced by all. The creative life of such a community rests on a belief in the ever continuing possibility of revealing still hidden truths. In *Science, Faith and Society*, I interpreted this as a belief in a spiritual reality, which, being real, will bear surprising fruit indefinitely. To-day I should prefer to call it a belief in the reality of emergent meaning and truth.

The mental pursuits of society depend for their resources and their protection on its economic and legal order. Consequently, the pursuit of profit and power will interact with the growth of thought in society. The extent of this interaction will vary among the different branches of thought. As for the effect on science, its progress can hardly be deflected at all from its intrinsic interests; it can only be stunted or stopped by an infringement of its autonomy.

This recognition of the symbiosis between thought and society brings us closer to the Marxist position and at the same time makes our difference from it clear. Marxism-Leninism denies the intrinsic creative powers of thought. Any claim to independence by scientists, scholars or artists must then appear as a plea for self-indulgence. A dedication to the pursuit of science, wherever it may lead, becomes disloyalty to the power responsible for the public welfare.

Since this power regards itself as the embodiment of historic destiny and as dispenser of history's promises to mankind, it can acknowledge no superior claims of truth, justice or morality. Alternatively, materialistic (or romantic) philosophies, denying any universal claims to the standards of truth, justice or morality, may deprive citizens of any grounds for appealing to these standards and thus endow the government with absolute power. The two processes are in fact fused in their joint justification of force as superior to mind.

But we must add here an additional process which makes violence the embodiment of the values it overrides. Those who in our day brought into power governments exempt from the standards of humanity were themselves prompted by an intense passion for the ideals which they so contemptuously brushed aside. They had rejected the overt professions of these ideals as philosophically unsound, hypocritical and specious, but they had covertly injected the same ideals into the new despotisms which they set up. Thus these ideals became immanent in the violence which ruthlessly rejected them. By virtue of this *moral inversion* (as I have later called it), the very immoralism of this power became a token of its moral purity. In view of its internal structure it could honestly reject any accusations of immorality in the very breath of proclaiming its own immorality.

A régime thus constituted claims to embody, besides morality, the ideals of justice, of the arts and sciences—in short all manner of truth. But here it overreaches itself. The rebellious movement which has transformed the régime of most Communist countries since Stalin's death was stirred up by seething demands for truth. I shall quote here from the writings of Nicolas Gimes, a Hungarian Communist who, though he had shortly before been a faithful Stalinist, turned against Stalinism in the Hungarian Revolution of October 1956. The following passage was published three weeks before the revolution.

> Slowly we had come to believe, at least with the greater, the dominant part of our consciousness . . . that there are two kinds of truth, that the truth of the Party and the people can be different and can be more important than the objective truth and that truth and political expediency are in fact identical. This is a terrible thought . . . if the criterion of truth is political expediency, then even a life can be 'true' . . . even a trumped up political trial can be 'true' . . . . And so we arrived at the outlook which infected not only those who thought up the faked political trials but often affected even the victims; the outlook which poisoned our whole public life, penetrated the remotest corners of our thinking, obscured our vision, paralysed our critical faculties and finally rendered many of us incapable of simply sensing or apprehending truth. This is how it was, it is no use denying it.

The author of these lines was executed in Budapest in 1958 at the orders of Moscow.

Since 1956, every successive report has made it clearer that the demand for truth is the motive force of renewal throughout the Soviet empire. It revives the great tradition of the intellectuals which originated in the Enlightenment. Marxist revisionism is an attempt to restore the original humanism of the Enlightenment and to stabilise it against the kind of self-destruction which led to Stalinism. Western writers have ascribed this movement of liberation to a higher level of industrialisation. They are still prisoners of the philosophic corruption which has plunged man's hopes into darkness. Nicolas Gimes and his comrades fought to redeem man's faith in truth from this corruption.

I have argued that a general respect for truth is all that is needed for society to be free. The way freedom and truth have proved identical in the battle against Stalinism bears out my views. I hope to see a modern theory of freedom, conceived on these lines, emerging from this battle.

OXFORD
December 1963

I

SCIENCE AND REALITY

I

What is the nature of science? Given any amount of experience, can scientific propositions be derived from it by the application of some explicit rules of procedure? Let us limit ourselves for the sake of simplicity to the exact sciences and conveniently assume that all relevant experience is given us in the form of numerical measurements; so that we are presented with a list of figures representing positions, masses, times, velocities, wavelengths, etc., from which we have to derive some mathematical law of nature. Could we do that by the application of definite operations? Certainly not. Granted for the sake of argument that we could discover somehow which of the figures can be connected so that one group determines the other; there would be an infinite number of mathematical functions available for the representation of the former in terms of the latter. There are many forms of mathematical series—such as power series, harmonic series, etc.—each of which can be used in an infinite variety of fashions to approximate the existing relationship between any given set of numerical data to any desired degree. Never yet has a definite rule been laid down by which any particular mathematical function can be recognized, among the infinite number of those offering themselves for choice, as the one which expresses a natural law. It is true that each of the infinite number of available functions will, in general, lead to a different prediction when applied to new observations, but this does not provide the requisite test for making a selection among them. If we pick out those which predict rightly, we still have an infinite number on our hands. The situation is in fact only changed by the addition of a few more data—namely, the 'predicted' data—to those from which we had originally started. We are not brought appreciably nearer towards definitely selecting any particular function from the infinite number of those available.

Now, I am not suggesting that it is impossible to find natural laws; but only that this is not done, and cannot be done, by applying some explicitly known operation to the given evidence of measurements. And to bring my argument a little closer to the actual experience of science, I shall now restate it as follows. We ask: Could a mathematical function connecting observable instrument readings ever constitute what we are accustomed to regard as a natural law in science? For example, if we were to state our knowledge concerning the path of a planet in these terms: 'That setting certain telescopes at certain angles at certain times a luminous disc of a certain size will be observed'—does that properly express a natural law of planetary motion? No: it is obvious that such a prediction is not equivalent to a proposition concerning planetary motion. Firstly, because we will in general be claiming too much and our prediction will prove often false even though the underlying proposition on planetary motion was correct: for a cloud may make the planet invisible to the eye, or else the soil may give way under the observatory, or some other of a hundred and one possible errors or obstacles may falsify observation or make it unworkable. Secondly, we would be claiming too little, since the presence of a planet at certain points of space—as postulated by its law of motion—may manifest itself in an indefinite variety of ways, the majority of which could not, on account of their sheer multitude, ever be explicitly predicted; and many of which may even be unthinkable to-day as they may be due to arise from yet unknown properties of matter or a host of other factors unknown at present, though inherent in our system.

There is, in fact, an essential feature lacking in both of the foregoing representations of science, which can be perhaps best pointed out by using yet a third picture of science. Suppose we wake up at night to the sound of a noise as of rummaging in a neighbouring unoccupied room. Is it the wind? A burglar? A rat? . . . We try to guess. Was that a foot-fall? That means a burglar! Convinced, we pluck up courage, rise, and proceed to verify our assumption.

Here are some of the features of a scientific discovery that we had missed before. The theory of the burglar—which represents our discovery—does not involve any definite relation of observational data from which further new observations can be definitely predicted. It is consistent with an infinite number of possible future observations. Yet the theory of the burglar is substantial and definite enough; it may even be capable of proof beyond any reasonable doubt in a court of law. In the light of common sense there is nothing curious in this: it merely makes it clear that the burglar is being assumed to be a real entity; a real burglar. So that we may even reverse this by saying that science is assuming something real whenever its propositions resemble the theory of the burglar. In this sense an assertion concerning the path of a planet may be said to be a proposition concerning something real, it being open to verification not only by some definite but also by many as yet quite undefined observations. We often hear of scientific theories gaining confirmation by later observations in a manner described as most surprising and audacious. The feat of Max v. Laue (1912) jointly confirming by the diffraction of X-rays in crystals both the wave nature of the X-rays and the lattice structure of crystals, is often praised as a striking feat of genius. It appears of the essence of scientific propositions that they are capable of bearing such distant and unexpected fruit; and we may conclude, therefore, that it is also of their essence to be concerned with reality.

A second significant feature of the discovery of the burglar, closely connected with what has just been said, is the way in which it is made. Curious noises are noticed; speculations about wind, rats, burglars, follow, and finally one more clue being noticed and taken to be decisive, the burglar theory is established. We see here a consistent effort at guessing—and at guessing right. The process starts with the very moment when, certain impressions being felt to be unusual and suggestive, a 'problem' is presenting itself to the mind; it continues with the collection of clues with an eye to a definite line of solving the problem; and it culminates in the guess of a definite solution.

But there is a difference between the solution offered by the burglar theory and that offered by a new scientific proposition. The first selects for its solution a known element of reality—namely burglars—the second often postulates an entirely new one. The vast growth of science in the last 300 years proves massively that new aspects of reality are constantly being added to those known before. Whence can we guess the presence of a real relationship between observed data, if its existence has never before been known?

We must go back to the process by which we usually first establish the reality of certain things around us. Our principal clue to the reality of an object is its possession of a coherent outline. It was the merit of Gestalt psychology to make us aware of the remarkable performance involved in perceiving shapes. Take, for example, a ball or an egg: we can see their shapes at a glance. Yet suppose that instead of the impression made on our eye by an aggregate of white points forming the surface of an egg, we were presented with another, logically equivalent, presentation of these points as given by a list of their spacial co-ordinate values. It would take years of labour to discover the shape inherent in this aggregate of figures—provided it could be guessed at all. The perception of the egg from the list of co-ordinate values would, in fact, be a feat rather similar in nature and measure of intellectual achievement to the discovery of the Copernican system. We can say, therefore, that the capacity of scientists to guess the presence of shapes as tokens of reality differs from the capacity of our ordinary perception, only by the fact that it can integrate shapes presented to it in terms which the perception of ordinary people cannot readily handle. The scientist's intuition can integrate widely dispersed data, camouflaged by sundry irrelevant connexions, and indeed seek out such data by experiments guided by a dim foreknowledge of the possibilities which lie ahead. These perceptions may be erroneous; just as the shape of a camouflaged body may be erroneously perceived in everyday life. I am concerned here only with showing that some of the characteristic features of the propositions of science exclude the possibility of deriving these by definite operations applied to primary observations; and to demonstrate that the process of their discovery must involve an intuitive perception of the real structure of natural phenomena. In the rest of this lecture I shall examine this position further and also point out (in section v) the necessity of amplifying it in some important respects.

II

However, would it not seem that our daily experience compels us with the force of logical necessity to accept certain natural laws as true? Generalizations such as 'all men must die' or 'the sun sheds daylight' seem to follow from experience without any intervention of an intuitive faculty on the part of ourselves as observers. But this only shows that we incline to regard our own particular convictions as inescapable. For these generalizations are quite commonly denied by primitive peoples. Such people believe that no man ever dies, except as a victim of evil magic, and some of them also believe that the sun crosses back by night to the east without shedding any light in its course. Their denial of natural death is part of their general belief that events which are harmful to man are never natural, but always the outcome of magic wrought by some malevolent person. In this magical interpretation of experience we see some causes which to us are massive and plain (such as a stone smashing a man's skull) regarded as incidental or even irrelevant to the event, while certain remote incidents (like the passing overhead of a rare bird) which to us appear to have no conceivable bearing on it are seized upon as its effective causes.

The primitive peoples holding these magical views are of normal intelligence. Yet they not only find their views wholly consistent with everyday experience, but will uphold them firmly in the face of any attempt on the part of Europeans to refute them by reference to such experience. For the terms of interpretation which we derive from our intuition of the fundamental nature of external reality cannot be readily proved inadequate by pointing at any particular new element of experience.

We are thus, it would seem, in danger of the opposite extreme: namely, of losing sight of any difference between the rival claims of the magical and the naturalistic interpretations of events. Now, it is true that there is a poetic truth expressed in primitive magical theory which is commonly found in our works of fiction. If a man in a novel is killed by accident, the event must have some human justification; the question of the Bridge of San Luis Rey can never be disregarded in a work of art. The naturalistic view of a man's death, say by a rail accident, robs human fate of some of its proper meaning; tending to reduce it to 'a tale told by an idiot, signifying nothing'. But at the same time the naturalistic view opens such a noble vista of the natural order of things which are inaccessible to the magical view, and establishes so much more decent and responsible relationships between human beings, that we must not hesitate to accept it as the truer of the two.

A similar competitive conflict comes into view in contrasting the medieval and the scientific outlook. It is usually overlooked that medieval catholic philosophy was first established in a world imbued with scientific rationalism. St. Augustine, who above all laid the foundations of catholic philosophy, testifies amply in his *Confessions* to his profound interest in science before his conversion. But as he approached conversion he came to regard all scientific knowledge as barren and its pursuit as spiritually misleading. The battle which round the year 380 was fought in Augustine's mind was won by his fervent desire for a certainty of God which he felt to be endangered by the intellectual pride of men pursuing the chain of second causes. 'Nor doest Thou draw near', he wrote, 'but to the contrite in heart, nor art found by the proud, no, not though by curious skill they could number the stars and the sand, and measure the starry heavens, and track the course of the planets' (*Conf.*, bk. v, p. 3).

Eleven hundred years later we see St. Augustine's spell broken in its turn by a gradual change in the balance of mental desires. The secular spirit, critical, extrovert, rationalist, spread into many other fields before it revived the scientific study of nature. Science was a late child of the Renaissance; in fact by the time of Copernicus' and Vesalius' discoveries, the Renaissance had passed its peak and was falling under the shadow of the Counter-Reformation. Both Copernicus and Vesalius discovered new facts *because* they abandoned established authority —and not the other way round. Copernicus was affected by the new spirit while studying canon law at Italian universities around the year 1500. He returned home from Italy where so-called Pythagorean doctrines were then freely discussed, in strong and irrevocable possession of the heliocentric view.[1] When Vesalius first examined the human heart and did not find the channel through the septum postulated by Galen, he assumed that it was invisible to the eye; but some years later, with his faith in authority shaken, he declared dramatically that it did not exist.

And I think that to-day we can feel the balance of mental needs tilting back once again. Science is not so emphatic any more in disregarding how far its generalizations make sense when extended to the world as a whole. It is doubtful whether to-day scientists would accept without murmur, as they still did at the end of the nineteenth century, a view like that of Laplace and Poincaré about the nature of the universe. Poincaré had shown that from Laplace's mechanical theory there followed that every phase of atomic configuration must go on recurring cyclically to infinity and that every conceivable configuration (of the same total energy) keeps recurring likewise—so that on revisiting our universe one day we may have a chance of finding ourselves going through life once more, but this time in the reverse direction starting with a revival of our dead bodies and ending our lives as babies, eventually to be absorbed by the maternal womb. To-day, I believe such manifestly absurd conclusions would be seriously held against a scientific system which ventured to put them forward. In fact the modern study of cosmogony has involved—as Sir Edmund Whittaker has pointed out in his Riddell Lectures of 1944—a renewal of interest in the universe as one comprehensive whole. Moreover, since the advent of relativity, scientists have become increasingly confident that natural laws can be discovered by a systematic elimination of unwarranted assumptions implied in our way of thinking and this has strengthened our sense of rationality in the universe.

[1] Agnes M. Clarke, *Enc. Brit.*, 14th ed., vol. vi, p. 400. E. A. Burtt in *The Metaphysical Foundations of Modern Science* makes it particularly clear that from the empirical point of view there was nothing to be said for the Copernican view at the time of its propounding. 'Contemporary empiricists', he says on p. 25, 'had they lived in the sixteenth century, would have been first to scoff out of court the new philosophy of the universe.'

We conclude that objective experience cannot compel a decision either between the magical and the naturalist interpretation of daily life or between the scientific and the theological interpretation of nature; it may favour one or the other, but the decision can be found only by a process of arbitration in which alternative forms of mental satisfaction will be weighed in the balance. The foundations of such decisions will be ascertained in my third lecture. Now I return to the analysis of science.

III

The part played by new observations and experiment in the process of discovery in science is usually over-estimated. The popular conception of the scientist patiently collecting observations, unprejudiced by any theory, until finally he succeeds in establishing a great new generalization, is quite false. 'Science advances in two ways,' remarks Jeans, 'by the discovery of new facts, and by the discovery of mechanisms or systems which account for the facts already known. The outstanding landmarks in the progress of science have all been of the second kind.' As examples he quotes the work of Copernicus, Newton, Darwin, and Einstein. We could add Dalton's atomic theory of chemical combination, de Broglie's wave theory of matter, Heisenberg's and Schrödinger's quantum-mechanics, Dirac's theory of the electron and positron. In a number of these discoveries predictions of the highest importance were involved which often came to light only years after the discovery was made. All this new knowledge of nature was acquired merely by the reconsideration of known phenomena in a new context which was felt to be more rational and more real.

The assumptions guiding these discoveries were the premisses of science, that is, the fundamental guesses of science concerning the nature of things. With these premisses I shall not deal in detail but only note that great discoveries achieved by the mere reconsideration of known phenomena are a striking illustration of the presence of these premisses and a mark of their rightness.[1]

It will be objected—following yet another widespread popular misconception—that even though scientists do occasionally put forward in advance of evidence assumptions that appear

[1] A brief discussion of these premisses is given in Appendix, 1.

a priori plausible to them, they only use them as a 'working hypothesis' and are ready immediately to abandon them in face of conflicting observational evidence. This, however, is either meaningless or untrue. If it means that a scientific proposition is abandoned whenever some new observation is accepted as evidence against it, then the statement is, of course, tautologous. If it suggests that any new observation which formally contradicts a proposition leads to its abandonment, it is, equally obviously, false. The periodic system of elements is formally contradicted by the fact that argon and potassium, as well as tellurium and iodine, fit in only in a sequence of decreasing, instead of increasing, atomic weights. This contradiction, however, did at no time cause the system to be abandoned. The quantum theory of light was first proposed by Einstein—and upheld subsequently for twenty years—in spite of its being in sharp conflict with the evidence of optical diffraction.[1]

This position is indeed to be expected on the grounds of our introductory analysis. We had established there that scientific propositions do not refer definitely to any observable facts but are like statements about the presence of a burglar next door—describing something real which may manifest itself in many indefinite ways. We have seen that there exist therefore no explicit rules by which a scientific proposition can be obtained from observational data, and we must therefore accept also that no explicit rules can exist to decide whether to uphold or abandon any scientific proposition in face of any particular new observation. The part of observation is to supply clues for the apprehension of reality: that is the process underlying scientific discovery. The apprehension of reality thus gained forms in its turn a clue to future observations: that is the process underlying verification. In both processes there is involved an intuition of the relation between observation and reality: a faculty which can range over all grades of sagacity, from the highest level present in the inspired guesses of scientific genius, down to a minimum required for ordinary perception. Verification, even though usually more subject to rules than discovery, rests ultimately on mental powers which go beyond the application of any definite rules.

Such a conclusion may appear less strange if we consider the

[1] Further discussion in Appendix, 2.

phases through which the propositions of science are usually brought into existence. In the course of any single experimental inquiry the mutual stimulus between intuition and observation goes on all the time and takes on the most varied forms. Most of the time is spent in fruitless efforts, sustained by a fascination which will take beating after beating for months on end, and produce ever new outbursts of hope, each as fresh as the last so bitterly crushed the week or month before. Vague shapes of the surmised truth suddenly take on the sharp outlines of certainty, only to dissolve again in the light of second thoughts or of further experimental observations. Yet from time to time certain visions of the truth, having made their appearance, continue to gain strength both by further reflection and additional evidence. These are the claims which may be accepted as final by the investigator and for which he may assume public responsibility by communicating them in print. This is how scientific propositions normally come into existence.

The certainty of such propositions can differ therefore only in degree from that of previous preliminary results, many of which had appeared final at first and only later turned out to have been only preliminary. Which is not to say that we must always remain in doubt, but only that our decision what to accept as finally established cannot be wholly derived from any explicit rules but must be taken in the light of our own personal judgement of the evidence.

Nor am I saying that there are no rules to guide verification, but only that there are none which can be relied on in the last resort. Take the most important rules of experimental verification: reproducibility of results; agreement between determinations made by different and independent methods; fulfilment of predictions. These are powerful criteria; but I could give you examples in which they were all fulfilled and yet the statement which they seemed to confirm later turned out to be false. The most striking agreement with experiment may occasionally be revealed later to be based on mere coincidence, as it was in these cases. Agreement with experiment will therefore always leave some conceivable doubt as to the truth of a proposition and it is for the scientist to judge whether he wants to set aside such doubt as unreasonable or not.[1]

[1] See Appendix, 3.

Similar considerations apply, of course, to the accepted rules of refutation. It is true enough that the scientist must be prepared to submit at any moment to the adverse verdict of observational evidence. But not blindly. That is what I have illustrated by the examples of the periodic system and the quantum theory of light, both upheld in spite of contradicting evidence. There is always the possibility that, as in these cases, a deviation may not affect the essential correctness of a proposition. The example of the periodic system and of the quantum theory of light both show how the objections raised by a contradiction to a theory may eventually be met not by abandoning it but rather by carrying it one step further: Any exception to a rule may thus conceivably involve, not its refutation, but its elucidation and hence the confirmation of its deeper meaning.

The process of explaining away deviations is in fact quite indispensable to the daily routine of research. In my laboratory I find the laws of nature formally contradicted at every hour, but I explain this away by the assumption of experimental error. I know that this may cause me one day to explain away a fundamentally new phenomenon and to miss a great discovery. Such things have often happened in the history of science. Yet I shall continue to explain away my odd results, for if every anomaly observed in my laboratory were taken at its face value, research would instantly degenerate into a wild-goose chase after imaginary fundamental novelties.

We may conclude that just as there is no proof of a proposition in natural science which cannot conceivably turn out to be incomplete, so also there is no refutation which cannot conceivably turn out to have been unfounded. There is a residue of personal judgement required in deciding—as the scientist eventually must—what weight to attach to any particular set of evidence in regard to the validity of a particular proposition.

IV

The propositions of science thus appear to be in the nature of guesses. They are founded on the assumptions of science concerning the structure of the universe and on the evidence of observations collected by the methods of science; they are subject to a process of verification in the light of further observations according to the rules of science; but their conjectural character remains inherent in them.

As I am convinced that there is great truth in science I do not consider its guesses as unfounded. Let me resume therefore my examination of this guesswork and see what method, if any, can be discovered in its operations.

In science the process of guessing starts when the novice feels first attracted to science and is then attracted further towards a certain field of problems. This guesswork involves the assessment of the young person's own yet largely undisclosed abilities, and of a scientific material, yet uncollected or even unobserved, to which he may later successfully apply his abilities. It involves the sensing of hidden gifts in himself and of hidden facts in nature, from which two, in combination, will spring one day his ideas that are to guide him to discovery. It is characteristic of the process of scientific conjecture that it can guess, as in this case, the several consecutive elements of a coherent sequence—even though each step guessed at a time can be justified only by the success of the further yet unguessed steps with which it will eventually combine to the final solution. This is particularly clear in the case of a mathematical discovery consisting of a whole new chain of arguments. In his book *How to Solve It*, G. Polya has compared such discovery with an arch where every stone depends for its stability on the presence of the others, and pointed out the paradox that the stones are in fact put in one at a time. The sequence of operations leading up to the chemical synthesis of an unknown body is in the same category; for unless final success is achieved, all the work is largely or entirely wasted. In order to guess a series of such steps, an intimation of approaching nearer towards a solution must be received at every step. There must be a sufficient foreknowledge of the whole solution to guide conjecture with reasonable probability in making the right choice at each consecutive stage. The process resembles the creation of a work of art which is firmly guided by a fundamental vision of the final whole, even though that whole can be definitely conceived only in terms of its yet undiscovered particulars—with the remarkable difference, however, that in natural science the final whole lies not within the powers of our shaping, but must give a true picture of a hidden pattern of the outer world.

I have previously suggested that the process of discovery is akin to the recognition of shapes as analysed by Gestalt psychology. Köhler assumes that the perception of shapes is caused by the spontaneous reorganization of the physical traces made by sense impressions inside our sense organs. He assumes that these traces somehow interact and coalesce to a dynamic order, the formation of which produces in the observer the perception of a shape. We may follow up our parallel between discovery and Gestalt perception by regarding the process of discovery as a spontaneous coalescence of the elements which must combine to its achievement. Potential discovery may be thought to attract the mind which will reveal it—inflaming the scientist with creative desire and imparting to him a foreknowledge of itself; guiding him from clue to clue and from surmise to surmise. The testing hand, the straining eye, the ransacked brain, may be thought to be all labouring under the common spell of a potential discovery striving to emerge into actuality.

The conditions in which discovery usually occurs and the general way of its happening certainly show it in fact to be a process of emergence rather than a feat of operative action. Operational skill, such as the facility for carrying out rapidly and accurately a large number of measurements and calculations counts for little in a scientist. There exist many excellent manuals on methods of computation and on every form of experimental technique. There are specifications for testing materials and rules for drawing up statistics. There are also manuals for triangulation and the drawing of exact maps. But there are no manuals prescribing the conduct of research; clearly because its method cannot be definitely set out. Only routine progress—such as the production of good maps and charts of all kinds—can be made by rules alone. The rules of research cannot usefully be codified at all. Like the rules of all other higher arts, they are embodied in practice alone. There is a popular belief that a procedure of empirical discovery has been revealed and established by Francis Bacon. But actually his prescription of making discoveries by collecting all the facts and passing them through an automatic mill was a travesty of research. The study of heuristics, i.e., the inquiry into the general method of solving problems in mathematics, has been recently revived by G. Polya in his *How to Solve It*. But his excellent little book only proves that discovery, far from representing a definite mental operation, is an extremely delicate and personal art which can be but little assisted by any formulated precepts.

There can actually be no doubt that, at any rate in mathematics, the most essential phase of discovery represents a process of spontaneous emergence. This was first described by Poincaré, who in *Science et Methode* has analysed the way some of his own great mathematical discoveries were made. He noted that discovery does not usually occur at the culmination of mental effort—the way you reach the peak of a mountain by putting in your last ounce of strength—but more often comes in a flash after a period of rest or distraction. Our labours are spent as it were in an unsuccessful scramble among the rocks and in the gullies on the flanks of the hill and then when we would give up for the moment and settle down to tea we suddenly find ourselves transported to the top. All the efforts of the discoverer are but preparations for the main event of discovery, which eventually takes place—if at all—by a process of spontaneous mental reorganization uncontrolled by conscious effort.

This outline of mathematical discovery has been confirmed by all subsequent writers and a similar rhythm has been observed over a wide field of other creative activities of the mind. The four phases observed in mathematical discovery, namely, Preparation, Incubation, Illumination, and Verification (as Wallas has called them) were found also in the course of discovery in natural science and they can be traced similarly through the process leading to the creation of a work of art. They are very clearly reproduced also in the mental effort leading to the recovery of a lost recollection. The solution of riddles, the invention of practical devices, the recognition of indistinct shapes, the diagnosis of an illness, the identification of a rare species, and many other forms of guessing right seem to conform to the same pattern. Among these I would include also the prayerful search for God. The report of St. Augustine of his long labours to achieve faith in Christianity, abruptly culminating in his conversion, which he immediately recognized as final and followed up by the lifelong vindication of the suddenly acquired faith, certainly reveals all the characteristic stages of the creative rhythm.

All these processes of creative guesswork have in common that they are guided by the urge to make contact with a reality, which is felt to be there already to start with, waiting to be apprehended. That is why the egg of Columbus is the proverbial symbol of great discovery. It suggests that great discovery is the realization of something obvious; a presence staring us in the face, waiting until we open our eyes.

In this light it may appear perhaps more appropriate to regard discovery in natural sciences as guided not so much by the potentiality of a scientific proposition as by an aspect of nature seeking realization in our minds. The process of scientific intuition is then brought into analogy with extra-sensory perception as established by Rhine (1934). It would appear particularly kindred to the acts of precognition or apparent clairvoyance, that is the guessing of objects not known to anyone. The intuitive phase of natural discovery and extra-sensory perception have it in common that they rely on an effort of mental concentration to evoke the knowledge of a real thing never seen before. There is ample evidence that, like extra-sensory perception, heuristic intuition works in a fairly determinate fashion. Two scientists faced with a similar set of facts will often hit on the same problem and discover the same solution to it. Coincident or nearly coincident discoveries by independent investigators are quite common and would be even more frequently observed but for the fact that rapid publication of an earlier successful piece of work often prevents the completion of others which would soon follow after. Therefore, when denying that discovery can ever be achieved by carrying out a set of definite operations we need not place the process altogether outside the laws of nature but may continue to regard its course as closely limited by the circumstances facing the investigator. (The factors lying outside the control of circumstance will be dealt with in Section v.)

But the study of extra-sensory perception may have further lessons for the understanding of intuition. One of the most curious coincidences in the history of science was the almost simultaneous discovery of quantum-mechanics by Heisenberg and Born in the form of matrices and by Schrödinger in the form of wave mechanics, for in this case the two claims were first considered as conflicting. The starting-points of the two theories and their presentations of the problem, their whole mathematical apparatus were different; and above all—as Schrödinger pointed out in his paper eventually establishing the mathematical identity of the two—their departure from classical mechanics lay in diametrically opposite directions. It seems most reasonable to describe this event by saying that both investigators had an intuitive perception of the same hidden reality present in nature, but that they drew different descriptions of it; so different that on comparing them they thought them to represent disparate objects. Actually, Dirac was soon to prove that both representations were considerably off the mark, as they were in conflict with relativity. When corrected for this shortcoming the formulation of quantum-mechanics was found to be once more transformed practically out of recognition. This seems to conform to the experience of extra-sensory perception. When the drawing of an object is sensed by telepathy or precognition there is no tendency to reproduce its physical outline independent of its meaning but on the contrary '. . . everything seems to happen [writes Mr. Whateley Carington[1]] much more as if those who scored hits had been told, "Draw a Hand" for example, [rather] than "Copy this drawing of a Hand". It is, as one might say, the "idea" or "content", or "meaning" of the original that gets over, not the form.' Thus we may think of Heisenberg and Schrödinger both penetrating to the same meaning but drawing different pictures of it; so different that they did not themselves recognize their identical meaning.

It is tempting to include in this picture also the fact, which I have heard mentioned with surprise among mathematicians, that when a problem which had appeared insoluble for a long time is finally solved, there are often discovered a series of solutions which appear to be quite independent of one another. This could be accounted for by assuming that intuition had sensed a reality of which these various solutions represent different descriptions or aspects. Again among mathematicians I have heard a series of discoveries by one person described as follows: The first discovery is like a solitary island in a border-

[1] *Telepathy*, p. 36.

less expanse of sea. Then a second and third island are discovered without any apparent connexion. But gradually it becomes clear that the waters are ebbing away in mass and leaving behind what were at first little isolated islands as the peaks of one great chain of mountains. That is precisely what one would expect to happen if intuition first sensed the fundamental chain of thought, i.e., the mountain range, and consciousness then proceeded to describe it little by little. Actually these unusual processes do not differ in essence from the ordinary event of a hidden chain of mathematical reasoning being discovered by a series of stepwise advances.

Lastly, I mention with some hesitation, but with the conviction that they must be at least tentatively considered in this context, the curious coincidences between theoretical and experimental discovery, of which some remarkable cases occurred in the last 20 years or so. In 1923 de Broglie suggested that electrons may possess wave nature and in 1925 Davisson and Germer, not knowing of this theory, made their first observations of the phenomenon soon after to be recognized as the diffraction of these waves. The prediction of the positive electron, which was implied in Dirac's relativistic quantum-mechanics of 1928, was confirmed by the discovery of the particle in 1932 by Anderson, who had no knowledge of Dirac's work. And we may add the prediction of the meson by Yukawa's theory of nuclear fields (1935) and its contemporaneous discovery in cosmic rays, finally established by Anderson (1938). Could it be that the same intuitive contact guided these alternative approaches to the same hidden reality?

Intuition is always imperfect. Different pictures of the same reality will be of unequal value and most of them will contain but a vague or excessively distorted form of the truth. We must also consider the possibility of completely erroneous shots in the dark. These are common enough in all forms of guesswork as well as in tests of extra-sensory perception. If the mind is uninformed by intuitive contact with reality, it is bound to place unreal and fruitless interpretations on the evidence before it. A passer-by called in from the street on chance to conduct scientific investigations would undoubtedly demonstrate this clearly enough.

But if science is but guesswork, why consider one guess

better than another? In other words, what, if any, is the basis for considering a proposition of science as valid? We shall answer this question in stages throughout the subsequent lectures. At the moment we are only claiming that whoever accepts natural science, or any part of it, as true, must recognize also our faculty to guess the nature of things in the outer world.

The two somewhat disparate formulations of discovery achieved up to this point—namely, (1) spontaneous organization of mind and clues to the realization of potential discovery and (2) extra-sensory perception of reality called into consciousness by the aid of relevant clues—would become identical if we were to assume that the ordinary perception of Gestalt includes a process of extra-sensory perception. That is, if sense impressions were normally accompanied by an extra-sensory transmission of the meaning to be attached to them. The uncertainty of the latter process, as observed in the usual tests of extra-sensory perception, could be taken to account for illusions and other interpretative errors. Such speculations may, however, appear premature in view of our yet too scanty knowledge of extra-sensory perception. So let us return once more to the closer analysis of scientific discovery.

V

We have yet to recognize an important element of all personal judgements affecting scientific statements. Viewed from outside as we described him the scientist may appear as a mere truth-finding machine steered by intuitive sensitivity. But this view takes no account of the curious fact that he is himself the ultimate judge of what he accepts as true. His brain labours to satisfy its own demands according to criteria applied by its own judgement. It is like a game of patience in which the player has discretion to apply the rules to each run as he thinks fit. Or, to vary the simile, the scientist appears acting here as detective, policeman, judge, and jury all rolled into one. He apprehends certain clues as suspect; formulates the charge and examines the evidence both for and against it, admitting or rejecting such parts of it as he thinks fit, and finally pronounces judgement. While all the time, far from being neutral at heart, he is himself passionately interested in the outcome of the procedure. He must be, for otherwise he will never discover a problem at all and certainly not advance towards its solution. '. . . To solve a serious scientific problem [writes Polya] will-power is needed that can outlast years of toil and bitter disappointments. . . .' 'We are elated when our forecast comes true. We are depressed when the way which we have followed with some confidence is suddenly blocked, and our determination wavers.' There is a strong temptation here to avoid discomfiture by paying insufficient attention to such evidence as obstructs our path. Starting from some intuitive preconception of the truth, and straining every nerve to prove this to be correct—it may be very difficult for the scientist not to overshoot the mark in trying to verify his suppositions. The Bible says: 'Correct a wise man and he will love you.' The scientist ought to be delighted when his theory, supported by a series of previous observations, appears to collapse in the light of his latest experiments. If he was wrong, then he has just escaped establishing a falsehood and been given a timely warning to turn in a new direction. But that is not how he feels. He is dejected and confused, and can only think of possible ways of explaining away the obstructive observation.

And of course there is always the possibility that this may in fact be just the right thing to do. This may be precisely one of those cases when one has to disregard exceptions to start with and leave them for later consideration. His emotion, born of an intuition which penetrates deeper than the day-to-day evidence, may be quite right, and his correct procedure may be to persevere in following its guidance, even against the apparent evidence.

I have said before that problems of this kind can be resolved by no established rule and that the decision to be taken is a matter for the scientist's personal judgement; we now see that this judgement has a moral aspect to it. We see higher interests conflicting with lower interests. That must involve questions of conviction and of faithfulness to an ideal; it makes the scientist's judgement a matter of conscience.

Faithfulness to the scientific ideals of care and honest self-criticism is, of course, indispensable even for the execution of the simplest jobs in the workshop of science. It is the first thing that a student is taught on being apprenticed to science. But, alas, many students only learn to be 'conscientious' in the sense of being pedantic and sceptical, which may be paralysing to all advance in research. Scientific conscience cannot be satisfied by the fulfilment of any rules, since all rules are subject to its own interpretation. To verify references, for example, is a matter of mere routine conscientiousness and not of the kind of conscience of which I am thinking here. But real scientific conscience is involved in judging how far other people's data can be relied upon and avoiding at the same time the dangers of either too little or too much caution. And similarly all the more difficult decisions to be taken in the pursuit of a scientific investigation and its subsequent publication and public defence, involve matters of conscience, each of which is a test for the scientist's sincerity and devotion to scientific ideals.

The scientist takes complete responsibility for every one of these actions and particularly for the claims which he puts forward. If his statements are confirmed by others, in whatever form and in whatever manner, even though quite unthought of at the time when he first propounded them, he will claim to have been right. And conversely, if his work is proved wrong he will feel that he has failed. He cannot plead to have observed the rules, or to have been misled by other investigators' evidence or his own collaborators', or that he could not at the time have made the tests which eventually disproved his thesis. Such reasons can serve to explain his error but they can never justify it—for he is bound to no explicit rules and is entitled to accept or reject any evidence at his own discretion. The scientist's task is not to observe any allegedly correct procedure but to get the right results. He has to establish contact, by whatever means, with the hidden reality of which he is predicating. His conscience must therefore give its ultimate assent always from a sense of having established that contact. And he will accept therefore the duty of committing himself on the strength of evidence which can, admittedly, never be complete; and trust that such a gamble, when based on the dictates of his scientific conscience, is in fact his competent function and his proper chance of making his contribution to science.

We can clearly distinguish in all these phases of discovery the two different personal elements which enter into every scientific judgement and make it possible for the scientist to be judge in his own case. Intuitive impulses keep arising in him stimulated by some of the evidence but conflicting with other parts of it. One half of his mind keeps putting forward new claims, the other half keeps opposing them. Both these parties are blind, as either of them left to itself would lead indefinitely astray. Unfettered intuitive speculation would lead to extravagant wishful conclusions; while rigorous fulfilment of any set of critical rules would completely paralyse discovery. The conflict can be resolved only through a judicial decision by a third party standing above the contestants. The third party in the scientist's mind which transcends both his creative impulses and his critical caution, is his scientific conscience. We recognize the note struck by conscience in the tone of personal responsibility in which the scientist declares his ultimate claims. This indicates the presence of a moral element in the foundations of science; and my next lecture will elaborate this in further detail.

II

AUTHORITY AND CONSCIENCE

WE have seen that the propositions embodied in natural science are not derived by any definite rule from the data of experience. They are first arrived at by a form of guessing based on premisses which are by no means inescapable and cannot even be clearly defined; after which they are verified by a process of observational hardening which always leaves play to the scientist's personal judgement. In every judgement of scientific validity there thus remains implied the supposition that we accept the premisses of science and that the scientist's conscience can be relied upon.

In my present lecture I shall try to expose the grounds on which the premisses of science are being held among scientists to-day and to show how the consciences of scientists are found to be rooted in the same grounds.

I

The premisses which underlie science fall into two classes. There are the general assumptions about the nature of everyday experience which constitute the naturalistic—as opposed to the magical, mythological, etc.—outlook. And then the more particular assumptions underlying the process of scientific discovery and its verification. Neither are inborn. The children of primitive natives whose parents are inveterately confirmed in their magical interpretation of things, can be brought up without difficulty to a naturalistic view of nature in the schools run by local missionaries. The reverse would no doubt be just as easy to achieve; and Europeans brought up to believe in an elaborate system of magic could be made as impervious to science as are primitive natives to-day. The naturalistic view held by scientists as by other modern men to-day has its origin in their primary education.

The premisses underlying a major intellectual process are never formulated and transmitted in the form of definite precepts. When children learn to think naturalistically they do not acquire any explicit knowledge of the principles of causation.

They learn to regard events in terms of what we call natural causes and by practising such interpretations day by day they are eventually confirmed in the premisses underlying them. Much of this happens already when the child learns to speak in a language which describes events in naturalistic terms, and the process of acquiring speech offers a good example for the principles by which the premisses of thought are in general transmitted from one generation to the next. Speech is learned by intelligent imitation of the adult. Each word must be noted in a number of contexts until its meaning is roughly grasped; it must then be read in books and used for some time in speech and writing under guidance of the example of adults in order that its most important shades of meaning be mastered. This training can be supplemented by precept, but imitative practice must always remain its main principle. The same is true of the process by which the elements of the higher arts are assimilated. Painting, music, etc., can be learned only by practice, guided by intelligent imitation. And this applies also to the art of scientific discovery.

The premisses of science are taught to-day roughly in three stages. School science imparts a facility in using scientific terms to indicate the established doctrine, the dead letter of science. The university tries to bring this knowledge to life by making the student realize its uncertainties and its eternally provisional nature, and giving him perhaps a glimpse of the dormant implications which may yet emerge from the established doctrine. It also imparts the beginnings of scientific judgement by teaching the practice of experimental proof and giving a first experience in routine research. But a full initiation into the premisses of science can be gained only by the few who possess the gifts for becoming independent scientists, and they usually achieve it only through close personal association with the intimate views and practice of a distinguished master. In the great schools of research are fostered the most vital premisses of scientific discovery. A master's daily labours will reveal these to the intelligent student and impart to him also some of the master's personal intuitions by which his work is guided. The way he chooses problems, selects a technique, reacts to new clues and to unforeseen difficulties, discusses other scientists' work, and keeps speculating all the time about a hundred possibilities which are never to materialize, may transmit a reflection at least of his essential visions. This is why so often great scientists follow great masters as apprentices. Rutherford's work bore the clear imprint of his apprenticeship under J. J. Thomson. And no less than four Nobel Laureates are found in turn among the personal pupils of Rutherford. Some forms of science, such as psycho-analysis, can hardly be transmitted by precept. Every psycho-analyst to-day has either been analysed by Freud or by another psycho-analyst who has been so analysed, etc. (Perhaps a modern version of the Apostolic Succession.) Research in the chemistry of carbohydrates in Britain has been almost entirely the work of four scientists, Purdy, Irvine, Haworth, and Hirst, who followed each other in single file as masters and pupils.

Any effort made to understand something must be sustained by the belief that there is something there that can be understood. Its effort to learn to speak is prompted in the child by the conviction that speech means something. Guided by its love and trust of its guardians, it perceives the light of reason in their eyes, voices, and bearing and feels instinctively attracted towards the source of this light. It is impelled to imitate—and to understand better as it imitates further—these expressive actions of its adult guides.

Apprenticeship to the higher arts, and to science in particular, is accepted and pursued on similar grounds. The future scientist is attracted by popular scientific literature or by schoolwork in science long before he can form any true idea of the nature of scientific research. The morsels of science which he picks up—even though often dry or else speciously varnished—instil in him the intimation of intellectual treasures and creative joys far beyond his ken. His intuitive realization of a great system of valid thought and of an endless path of discovery sustain him in laboriously accumulating knowledge and urge him on to penetrate into intricate brain-racking theories. Sometimes he will also find a master whose work he admires and whose manner and outlook he accepts for his guidance. Thus his mind will become assimilated to the premisses of science. The scientific intuition of reality henceforth shapes his perception. He learns the methods of scientific investigation and accepts the standards of scientific value.

At every stage of his progress towards this end he is urged on by the belief that certain things as yet beyond his knowledge and even understanding are on the whole true and valuable, so that it is worth spending his most intensive efforts on mastering them. This represents a recognition of the authority of that which he is going to learn and of those from whom he is going to learn it. It is the same attitude as that of the child listening to its mother's voice and absorbing the meaning of speech. Both are based on an implicit belief in the significance and truth of the context which the learner is trying to master. A child could never learn to speak if it assumed that the words which are used in its hearing are meaningless; or even if it assumed that five out of ten words so used are meaningless. And similarly no one can become a scientist unless he presumes that the scientific doctrine and method are fundamentally sound and that their ultimate premisses can be unquestioningly accepted. We have here an instance of the process described epigrammatically by the Christian Church Fathers in the words: *fides quaerens intellectum*, faith in search of understanding.

An essential part is played in the process of learning, by a form of intelligent guessing similar to that which underlies the process of discovery. To assimilate the hidden premisses of a major artistic or intellectual process is in fact a minor feat of discovery. To understand science is to penetrate to the reality described by science; it represents an intuition of reality, for which the established practice and doctrine of science serve as clues. Apprenticeship in science may be regarded as a much simplified repetition of the whole series of discoveries by which the existing body of science was originally established.

Thus the authority to which the student of science submits tends to eliminate its own functions by establishing direct contact between the student and the reality of nature. As he approaches maturity the student will rely for his beliefs less and less on authority and more and more on his own judgement. His own intuition and conscience will take over responsibility in the measure in which authority is eclipsed. This does not mean that he will rely no more on the report of other scientists —far from it—but it means that such reliance will henceforth be entirely subject to his own judgement. Submission to authority will henceforth form merely a part of the process of discovery, for which—as for the process as a whole—he will assume full responsibility before his own conscience.

It follows that his teachers' personal views will never—or should never—be accepted by the pupil except as an embodiment of the general premisses of science. Students should be trained to share the ground on which their teachers stand and to take on this their stand for their own independence. The student will therefore practice a measure of criticism even during his period of study, and the teacher will gladly foster any signs of originality on the part of the student. But this must remain within proper limits; the process of learning must rely in the main on the acceptance of authority. Where necessary this acceptance must be enforced by discipline.

Naturally, there is here a field of possible conflicts between masters and pupils. The student who, on obtaining in the course of elementary practice an erroneous result from his chemical analysis, would claim to have made a fundamental discovery, would make no progress. He must be reprimanded and if necessary removed. But masters who try to impose their personal fads on their research students and (as I have known in one case) put pressure on them to confirm their theories, must be even more firmly opposed.

This kind of conflict is one among a number of kindred types which can occur in scientific life. We shall refer to others later on. If extreme conflicts between masters and pupils were widespread, the transmission of the premisses of science from one generation to the next would be impossible and science would soon become extinct. The continued existence of science is an expression of the fact that such conflicts are rare. They are so rare because masters and pupils do possess in general sufficiently sincere attachment to science and a sufficiently authentic vision of it to find therein a common ground for agreement. Their consciences on which they have ultimately to rely for guidance harmonize sufficiently to keep them in concord. Naturally, some masters may be uninspired, pedantic, and oppressive, others perhaps misguided by their personal bias. Some students may refuse to be led before even having mastered the elements of their subject. But these failings are so infrequent that the resulting occasional breaches can be settled without difficulty by appeal to general scientific opinion. The scandal is eliminated by conciliation or disciplinary measures, or it is at least isolated and allowed to burn out without much harm done.

Here, as in many other cases, ultimate adjustments in the process of transmitting the premisses of science depend on a well functioning scientific opinion—the discussion of which will allow us to penetrate further into the question why scientists usually agree so well among themselves.

II

The master-pupil relation is but an instance and a facet of a wider set of institutions, providing for mutual reliance and mutual discipline among scientists, by which the practice of discovery is ordered and the premisses of science are fostered and developed. I shall roughly outline the framework of these institutions.

In material terms the domain of science consists of certain periodicals and books, of research grants and salaries, of the buildings used for teaching and research. This domain is administered by scientists at whose disposal the requisite funds are placed from sources outside the world of science. Their administration consists, as we shall see, mainly in keeping up the standards of science and in providing opportunities for its spontaneous progress.

Let us consider this administration.

Take first the periodicals. No proposed contribution to science has a chance of becoming generally known unless it is published in print; and its chances of recognition are very poor unless it is published in one of the leading scientific journals. The referees and editors of these journals are responsible for excluding all matter which they consider unsound or irrelevant. They are charged with guarding a minimum standard for all published scientific literature.

On its publication a paper is laid open to scrutiny by all scientists who will proceed to form, and possibly also to express, an opinion on its value. They may doubt or altogether reject its claims, while its author will probably defend them. After a time a more or less settled opinion will prevail.

The third stage of public scrutiny through which a contribution to science must pass in order to become generally known and established is its incorporation in text-books or at least standard books of reference. This accords it the final seal of scientific authority and accredits it for teaching at universities and schools, as well as for popular dissemination to a wider public. Text-books are usually composed or at least edited by authoritative scientists and their general acceptance is in any case controlled by reviewers and teachers holding authority among scientists.

Next we come to scientific posts. Science is being actively pursued to-day mainly in endowed institutions, where scientists, on gaining a senior position, are allowed to use freely their own time and the grants and assistance assigned to them to pursue their own researches. This independence granted to mature scientists represents the very core of scientific life. It leaves all the initiative for the starting of new lines of research to the sovereign judgement of individual scientists. But the appointments to the posts granting this privilege must be controlled the more rigorously. The selection of scientific personnel depends largely on the value attached by scientific opinion to the different candidates' published work; in addition to which the advice of authoritative scientists is solicited in connexion with every important scientific appointment. The allocation of special research grants and the conferment of scientific degrees and distinctions is conducted on similar lines.

The establishment of opportunities for research in the form of buildings, laboratories, research funds, and salaries is also fashioned (within the limits of the total available resources) in accordance with the advice of scientists. They will try to assure a maximum rate of progress of science as a whole by allocating resources to the most active growing points of science.

Authority is not equally distributed among scientists. There is a hierarchy of influence; but exceptional authority is attached not so much to offices as to persons. A scientist is granted exceptional influence by the fact that his opinion is valued and asked for. He may then be elected on administrative committees, but this is not essential. The self-government of science is largely unofficial; the decisions lie with scientific opinion at large, focused and expressed on each particular occasion by the most competent experts commanding wide confidence. The maintenance of the same minimum standards all over the realm of science requires the ability to compare scientific merit in different fields. It is essential for this purpose that scientists shall appreciate not only work done in their own field but also to some extent that done in neighbouring fields; at least to the extent that they should know whom to consult in this respect and be able to form a critical estimate of the opinions thus obtained. This coherence of valuations throughout the whole range of science underlies the unity of science. It means that any statement recognized as valid in one part of science can, in general, be considered as underwritten by all scientists. It also results in a general homogeneity of and a mutual respect between all kinds of scientists, by virtue of which science forms an organic entity.

The government of science which I have briefly outlined here exercises no specific direction on the activities under its control. Its function is not to initiate but to grant or to withhold opportunity for research, publication, and teaching, to endorse or discredit contributions put forward by individuals. Yet this government is indispensable to the continued existence of science. Let us briefly survey what its operations amount to.

In the previous lecture I have examined scientific validity and took it to be the characteristic feature of science. But validity is by no means the only standard by which a scientific proposition is accepted or rejected. For example, an accurate determination of the speed at which water flows in the gutter at a particular moment of time is not a contribution to science. All parts of science must have some bearing on the system of science and also at least in some way be interesting in themselves, either for contemplation or practice. These three: validity, profundity, and intrinsic human interest underlie jointly the valuation of scientific results.

Suppose now for a moment that no limitations of value were imposed on the publication of scientific contributions in journals. The selection—which is indispensable in view of the limited space—would then have to be done by some neutral method—say drawing lots. Immediately the journals would be flooded with rubbish and valuable work would be crowded out and banished to obscurity. Cranks are always abounding who will send in spates of nonsense. Immature, confused, fantastic, or else plodding, pedestrian, irrelevant material would be pouring in. Swindlers and bunglers combining all variants of deception and self-deception would seek publicity. Buried among so much that is specious or slipshod, the few remaining valuable publications could hardly have a chance of being recognized. The swift and reliable contacts by which scientists to-day keep each other informed would be broken; they would be isolated and their mutual reliance and co-operation paralysed.

We need hardly go into this much further. Unless it is somehow assured that professional teachers and research workers will not lack scientific qualifications of a certain grade, the whole system of endowed scientific institutions is bound to dissolve in chaos and corruption. The experience of undeveloped countries where scientific opinion is imperfectly organized, teaches us that even a comparatively slight weakening of scientific control can have marked deleterious effects on the integrity and effectiveness of scientific activities.

It seems clear enough then that the self-governing institutions of science are effective in safeguarding the organized practice of science which embodies and transmits its premisses. But their functions are mainly protective and regulative and are themselves based, as we shall show in a moment, on the pre-existence of a general harmony of views among scientists. We shall get therefore nearer to the real basis of scientific life if we now focus our attention directly on the fact that scientists tend to agree so well with one another.

III

The consensus prevailing in modern science is certainly remarkable. Consider the fact that each scientist follows his own personal judgement for believing any particular claim of science and each is responsible for finding a problem and pursuing it in his own way; and that each again verifies and propounds his own results according to his personal judgement. Consider moreover that discovery is constantly at work, profoundly remoulding science in each generation. And yet in spite of such extreme individualism acting in so many widely disparate branches, and in spite of the general flux in which they are all involved, we see scientists continuing to agree on most points of science. Even though controversy never ceases among them, there is hardly a question on which they do not agree after a few years discussion.

The harmony between the views independently held by individual scientists shows itself also in the way they conduct the affairs of science. We have seen that there is no central authority exercising power over scientific life. It is all done at a multitude of dispersed points at the recommendation of a few scientists who happen either to be officially involved or drawn in as referees for the occasion. And yet in general such decisions do not clash but, on the contrary, can rely on wide approval. Two scientists acting unknown to each other as referees for the publication of one paper usually agree about its approximate value. Two referees reporting independently on an application for a higher degree rarely diverge greatly. Hundreds of published scientific papers pass review of thousands of scientific readers before any of them finds reason to protest against the insufficient standard of a paper. Among over four hundred Fellows of the Royal Society there are few who strike any of their scientific colleagues as clearly unworthy of the honour; nor have I yet heard bitter complaints that the claims of others to gain election have been scandalously neglected. The same would be found in respect to professors and holders of other positions of equal rank in universities.

The fundamental unanimity prevailing among scientists manifests itself—paradoxically perhaps—most clearly in the case of conflict. Every scientist feels the urge to convince his fellow scientists of the rightness of his own claims. Even though he may not succeed in that for the moment he would feel confident of achieving it sooner or later. It is only towards scientists that he feels that way. He does not mind what musicians think of his claims nor does he expect ever to convince them that he is right. The concern with the opinion of scientists and his belief that they are bound eventually to recognize the truth expresses his conviction that his mind and theirs operate from the same premisses. He is disturbed by the fact that the evidence which convinces him should fail to convince them, and feels that it must do so in the end.

However revolutionary the claims of a scientist may be—as were those inaugurated in our time by the discoverers of relativity, psycho-analysis, quantum mechanics, or extra-sensory perception—he will always meet any opposition of scientific opinion as it *is* by appealing against it to scientific opinion as he thinks it *ought to be*. Even though the new discovery may involve, as it did in the cases just mentioned, a reconsideration of the traditional grounds of science, the pioneer would still appeal to that tradition as the common ground between himself and his opponents; and they in their turn would always accept this premiss. They would accept also in particular the pioneer's reference to the example of earlier pioneers; to the struggle of Pasteur, Semmelweiss, Lister, Arhenius, van 't Hoff, and the rest, who had to brave the scientific opinion of their own times. It is part of the scientific tradition to be constantly on our guard against suppressing by mistake some great discovery, the claims of which may at first appear nonsensical on account of their novelty. Thus even in the most profound divisions that have yet occurred in science, the rebels and conservatives have alike remained firmly rooted in the same grounds. Accordingly, these conflicts have always been settled after a comparatively short time in a fashion which has proved acceptable to all scientists.

The origin of the spontaneous coherence prevailing among scientists is thus becoming clear. They are speaking with one voice because they are informed by the same tradition. We can see here the wider relationship, upholding and transmitting the premisses of science, of which the master-pupil relationship forms one facet. It consists in the whole system of scientific life rooted in a common tradition. Here is the ground on which the premisses of science are established; they are embodied in a tradition, the tradition of science.

The continued existence of science is an expression of the fact that scientists are agreed in accepting one tradition, and that all trust each other to be informed by this tradition. Suppose scientists were in the habit of regarding most of their fellows as cranks or charlatans. Fruitful discussion between them would become impossible and they would no more rely on each other's results nor act on each other's opinion. Thus their mutual collaboration on which scientific progress depends would be cut off. The processes of publication, of compiling text-books, of teaching juniors, of making appointments, and establishing new scientific institutions, would henceforth depend on the mere chance of who happened to make the decision. It would then become impossible to recognize any statement as a scientific proposition or to describe anyone as a scientist. Science would become practically extinct.

Nor could the coherence of scientific opinion be restored by the establishment of any kind of central authority. Supposing the President of the Royal Society were empowered to decide in the last resort every scientific question. The very large majority of his decisions would of course have no scientific value. All progress would stop. No recruit with any love of science would join any institution governed by such decisions. We see signs of such an influence even in ordinary well-run Government departments or other large-scale organizations, where administrative superiors allocate research tasks to mature scientists serving under their direction. It is a great sacrifice to anyone who loves discovery to join such an organization. And if the superiors were to impose their specific views on their subordinates, as they sometimes tend to do, the position of the subordinate would become altogether unbearable.

Nor can science be successfully guided by scientific opinion unless it is strictly understood that this opinion represents only a temporary and imperfect embodiment of the traditional standards of science. The scientist seeking guidance from scientific opinion must not be tempted to canvass primarily his fellow scientists' approval. Though his income, his independence, his influence, in fact his whole standing in the world will depend throughout his career on the amount of credit he can gain in the eyes of scientific opinion, he must not aim primarily at this credit, but only at satisfying the standards of science. For the shorter way of gaining credit with scientific opinion may lead far astray from good science. The quickest impression on the scientific world may be made not by publishing the whole truth and nothing but the truth, but rather by serving up an interesting and plausible story composed of parts of the truth with a little straight invention admixed to it. Such a composition, if judiciously guarded by interspersed ambiguities, will be extremely difficult to controvert, and in a field in which experiments are laborious or intrinsically difficult to reproduce may stand for years unchallenged. A considerable reputation can be built up and a very comfortable university post be gained before this kind of swindle transpires—if it ever does. If each scientist set to work every morning with the intention of doing the best bit of safe charlatanry which would just help him into a good post, there would soon exist no effective standards by which such deception could be detected. A community of scientists in which each would act only with an eye to please scientific opinion would find no scientific opinion to please. Only if scientists remain loyal to scientific ideals rather than try to achieve success with their fellow scientists can they form a community which will uphold these ideals. The discipline required to regulate the activities of scientists cannot be maintained by mere conformity to the actual demands of scientific opinion, but requires the support of moral conviction, stemming from devotion to science and prepared to operate independently of existing scientific opinion.

There is, naturally, always some compulsion involved in upholding order in science. The material domain of science, its journals and text-books, its research grants, laboratories, lecture rooms, and salaried positions, are granted for use and support on definite occasions and legally protected from use or interference by unauthorized persons. The conduct of teaching in universities and the administration of research laboratories involve the use of extensive compulsory powers. But the creative order of the scientific community is not the resultant of a clash between sheer organized force on the one hand and individuals pursuing their mere personal ends on the other. Scientists must feel under obligation to uphold the ideals of science and be guided by this obligation, both in exercising authority and in submitting to that of their fellows, otherwise science must die.

It would thus appear that when the premisses of science are held in common by the scientific community each must subscribe to them by an act of devotion. These premisses form not merely a guide to intuition, but also a guide to conscience; they are not merely indicative, but also normative. The tradition of science, it would seem, must be upheld as an unconditional demand if it is to be upheld at all. It can be made use of by scientists only if they place themselves at its service. It is a spiritual reality which stands over them and compels their allegiance.

I have spoken before of scientific conscience, as the normative principle arbitrating between intuitive impulses and critical procedure, and as the ultimate arbiter in the relationship between master and pupil. We see now how a scientific community organizes the conscience of its members through the joint cultivation of scientific ideals.

We may recall the various phases by which the scientist normally performs his emotional and moral surrender to science. The first approach of the youthful mind to science is prompted by a love of science and a faith in its great significance which precedes any real understanding of it. This primary surrender to the intellectual authority of science is indispensable to any serious effort of assimilating science. As a next step the youth aspiring to become a scientist will have to accept the example of great scientists, some living and many dead, and seek to derive from it an inspiration for his own future career. In many cases he will join a master and give him freely his admiration and trust. And presently, when actively engaging in the pursuit of discovery and passionately absorbed in solving a problem, he must strive against self-deception and for a true feeling of reality, even though he may be sorely tempted to be content with a less authentic satisfaction. Before claiming discovery he must listen to his scientific conscience. As he advances in life his professional conscience acquires a variety of new functions; in publishing papers, in criticizing those by other authors, in lecturing to students, in selecting candidates for appointments, in a hundred ways he has to form judgements that are ultimately guided by the ideal of science as interpreted by his conscience. Finally as a partner in the administration of science he fosters the spontaneous growth of science by extending his love and solicitude to every new original effort; thus again surrendering to the reality and inherent purpose of science.

The sharing of these various surrenders by all the members of the community of scientists undoubtedly adds to their strength. The knowledge that the same obligations to scientific ideals are generally accepted by all scientists effectively confirms their faith in the reality of these ideals. When each scientist largely relies for his views and information on the work of many others, and is prepared to vouch for their reliability before his own conscience, then the conscience of each is borne out by that of many others. There exists then a community of consciences jointly rooted in the same ideals recognized by all. And the community becomes an embodiment of these ideals and a living demonstration of their reality.

IV

The art of scientific work is so extensive and manifold that it can be passed on from one generation to the next only by a large number of specialists, each of whom fosters one particular branch of it. Therefore science can exist and continue to exist only because its premisses can be embodied in a tradition which can be held in common by a community. This is true also of all complex creative activities which are carried on beyond the lifetime of individuals. We may think for example of the law and of the Protestant Christian religion. Their continued life is based on traditions of a structure similar to that of science and it will help us to understand tradition in science—and also prepare us for the more general problems of society with which we want to deal later—if we proceed now to include such fields as law and religion in our further discussion.

We have seen how science is constantly revolutionized and perfected by its pioneers, while remaining firmly rooted in its tradition. Each generation of scientists applies, renews, and confirms scientific tradition in the light of their particular inspiration. Similarly we see judges deriving from past judicial practice the principles of the law and applying these creatively in the light of their conscience to ever new situations; and see how in doing so they revise in many particulars the very practice from which they derived their principles. Similarly to the Protestant the Bible serves as a creative tradition to be upheld and reinterpreted in new situations in the light of his conscience. While the Bible is held by him to mediate to the individual the revelation which it records, belief in this revelation is held to acquire the full value of faith only when it is affirmed by the individual's conscience. Conscience can then be used even to oppose the authority of the Bible where the Bible is found spiritually weak.

Such processes of creative renewal always imply an appeal from a tradition as it *is* to a tradition as it *ought to be*. That is to a spiritual reality embodied in tradition and transcending it. It expresses a belief in this superior reality and offers devotion to its service. We have seen how in science this devotion is first established at the stage of apprenticeship and we could parallel this act of initiation and dedication in the field of law or religion. But the similarity of these several collective activities of the mind dedicated to the cultivation of their respective traditions seem clearly enough established.

The realms of science, of law, and of Protestant religion which I have taken as examples of modern cultural communities are each subject to control by their own body of opinion. Scientific opinion, legal theory, Protestant theology are all formed by the consensus of independent individuals, rooted in a common tradition. In law and in religion, it is true, there prevails a measure of official doctrinal compulsion from a centre, which is almost entirely absent from science. The difference is marked; yet in spite of such compulsion as legal and religious life are subjected to, the conscience of the judge and of the minister bears an important responsibility in acting as its own interpreter of the law or of the Christian faith. Thus the life of science, the law, and the Protestant Church all three stand in contrast to the constitution, say, of the Catholic Church which denies to the believer's conscience the right to interpret the Christian dogma and reserves the final decision in such matters to his confessor. There is here the profound difference between two types of authority; one laying down general presuppositions, the other imposing *conclusions*. We may call the first a General, the latter a Specific Authority.

The difference between the two types of authority is decisive. It is illustrated by my earlier fictive supposition of the President of the Royal Society imposing specific conclusions on all scientists. The establishment of an authority of a specific type over science would be as destructive of science as the General Authority normally exercised by scientific opinion is indispensable to its continued existence. A closer analysis of the difference between the two types of authority will throw further light on the relation between authority and conscience —both in science and in other fields.

In my first lecture I have distinguished—not in so many words, but still quite clearly—between two kinds of rules. I have said, for example, that there are no strict rules by which a true scientific proposition could be discovered and demonstrated to be true; but that this can be done by the light of certain vague rules embodied in the art of scientific research. I showed that even though some of these rules—which should be regarded as rules of art—are very rigid, they always leave a significant margin, and sometimes considerable play, to personal judgement. Strict rules, like those of the multiplication table, on the other hand, leave practically no room for interpretation. The two kinds shade imperceptibly into one another, but that does not invalidate the distinction between them.

Being incapable of precise formulation, rules of art can be transmitted only by teaching the practice which embodies them. For major realms of creative thought this involves the passage of a tradition by each generation to the next. Every time this happens there is a possibility that the rule of art be subjected to a significant measure of reinterpretation and it is important to realize clearly what this involves.

How can we ever interpret a rule? By another rule? There can be only a finite number of tiers of rules so that such a regression would soon be exhausted. Let us assume then that all existing rules were united into one single code. Such a code of rules could obviously not contain prescriptions for its own reinterpretation.

It follows that every process of reinterpretation introduces elements which are wholly novel; and hence also that a traditional process of creative thought cannot be carried on without wholly new additions being made to existing tradition at every stage of transmission. In other words, it is logically impossible for tradition to operate without the addition of wholly original interpretative judgements at every stage of transmission.

To illustrate this, take the fields of law, religion, politics, manners, etc. There are of course numberless routine decisions to be taken at every hour which can be arrived at without any significant innovation. But there are always borderline cases requiring a measure of discretion, and even in routine cases there will often be an element of finer discrimination involved where a personal judgement is indispensable. The major principles of science, law, religion, etc., are continuously remoulded by decisions made in borderline cases and by the touch of personal judgement entering into almost every decision. And apart from this silent revolution steadily remoulding our heritage, there are the massive innovations introduced by the great pioneers. Yet each of these actions forms an essential part of the process of carrying on a tradition.

The main contrast between a régime of General Authority such as prevails in science, the law, etc., and the rule of a Specific Authority as constituted by the Catholic Church lies in the fact that the former leaves the decisions for interpreting traditional rules in the hands of numerous independent individuals while the latter centralizes such decisions at headquarters. A General Authority relies for the initiative in the gradual transformation of tradition on the intuitive impulses of the individual adherents of the community and it relies on their consciences to control their intuitions. The General Authority itself is but a more or less organized expression of the general opinion—scientific, legal, or religious—formed by the merging and interplay of all these individual contributions. Such a régime assumes that individual members are capable of making genuine contact with the reality underlying the existing tradition and of adding new and authentic interpretations to it. Innovation in this case is done at numerous growing points dispersed through the community, each of which may take the lead over the whole at any particular moment. A Specific Authority on the other hand makes all important reinterpretations and innovations by pronouncements from the centre. This centre alone is thought to have authentic contacts with the fundamental sources from which the existing tradition springs and can be renewed. Specific Authority demands therefore not only devotion to the tenets of a tradition but subordination of everyone's ultimate judgement to discretionary decision by an official centre.

We see emerging here two entirely different conceptions of authority, one demanding freedom where the other demands obedience. The contrast is important for the wider problems of society to which the third lecture will lead us.

Meanwhile let us attend further to the position of tradition under a General Authority. The freedom that we have postulated, for each generation to interpret the common heritage at its own discretion, may seem altogether disruptive. How can we speak of tradition as a firm ground on which, for example, the premisses of science rest and as the soil in which the consciences of scientists are rooted if tradition can be chopped and changed by a group of people who happen to be calling themselves scientists at a particular moment—and made by them into anything they are pleased to decree? Even though we admit that scientists (or lawyers or ministers) who have originally been initiated and dedicated to an existing body of tradition are not likely to turn it wilfully into a travesty of itself, the fact remains that new problems are constantly bound to arise—as, for example, in science to-day the claims of extra-sensory perception or the conflict between free research and national security—which one generation of scientists must decide with lasting effects on the tradition of science, acting entirely on its own responsibility. Are there no safeguards against such arbitrariness? And in any case, what validity can we ascribe to judgements made in this fashion?

I reply that it is impossible to safeguard against the mistakes of such decisions, because any authority established for such purpose would destroy science. It is in the nature of science that it can live only if individual scientists are regarded as competent to state their views and the consensus of their opinions is regarded as competent to decide all questions for science as a whole. In this sense the decisions of scientific opinion in scientific matters are always of right provided only they are sincere; and the scientists of any particular period are rightfully absolute masters under their conscience of the heritage of science. They will not decide without listening to one another's views and occasionally even to those of the wider public; they will also recall the lessons of the past and scientists in one region will try to learn from others elsewhere; they will weigh their decisions in regard to their future consequences—but both this procedure and the conclusions to be drawn therefrom will be for themselves to decide. Such insight as is vouchsafed to them when acting in the full sense of their responsibility to science represents their final portion of grace and to act on it represents their whole duty. Their decisions are inherently sovereign, because it is in the nature of science that no authority is conceivable which could competently overrule their verdict.

This does not mean that scientific opinion is inherently infallible. No; scientists will always make plenty of mistakes, which will become apparent in retrospect later. It is easy to see to-day, for example, how great pioneers like Julius Robert Mayer, Semmelweiss, or Pasteur were neglected and the success of their discoveries delayed. It is easy to distinguish among past periods of science, some, like the seventeenth century, which were more richly inspired, from others, like part of the eighteenth, which were almost stagnant by comparison. The styles of science can be compared in different regions and observed to incline towards pedantry here and excessive laxity elsewhere. There is infinite room both for contemporary criticism and later heart-searchings; but that does not impair the competent character of the actions subjected to such criticism. Rightful decisions may often turn out to be erroneous yet they remain rightful all the same.

To accord competence to the decisions of scientific opinion would, of course, be meaningless unless we ourselves accept science to be as a whole true and significant. We may accord the same competence to legal opinion and also to certain bodies of religious opinion, but probably not to astrological or fundamentalist opinion. If we believe in science we will accept competent scientific opinion as on the whole valid, even though the final validation of any proposition will always involve a fractional amount of personal responsibility on our own part.

Here are, so far, the final grounds on which the scientist holds his premisses and bases the decisions of his conscience, and on which he, and also others who believe in science, accept the decisions of scientists as competent and their views as on the whole valid. They consist in the acceptance of science itself as valid. I have given no reason yet why the scientist or anybody else should believe in science as a whole and not in astrology or fundamentalism. The scientist's conviction that science *works* is no better, so far, than the astrologist's belief in horoscopes or the fundamentalist's belief in the letter of the Bible. A belief always works in the eyes of the believer.

In the next lecture I shall try to find the grounds on which the decision is found between rival interpretations of nature. Such choices must, of course, be taken on wider premisses than those of science, though they must include these as one set of possible assumptions among many others. We may expect these wider premisses to sustain a wider intellectual life which includes the scientific world as one of its sections. In fact we can hardly expect it to comprise less than the entire intellectual life of society. We shall not be able to examine such a large field in any detail. But there is one feature which, judging from the internal life of science, we may expect to be essential to it. That is freedom. If the way in which truth is found *in* science is any guide as to how truth is to be found *about* science, the society in which this process can be properly conducted must be based on freedom; discussion about science must be free. In order to discover the conditions for maintaining such freedom, we shall start the next lecture by inquiring further into the manner in which freedom is maintained within science itself.

III
DEDICATION OR SERVITUDE

I

FREEDOM bears an old question-mark across its face. To prevent lawless conflict a paramount power is required: how can this power be prevented from suppressing freedom? How can it indeed fail to suppress it if it is to eliminate lawless strife? Government appears as essentially supreme and absolute, leaving no room for freedom.

But we have said that in the world of science, which is an organized social body, there is freedom and that freedom is even essential to the maintenance of its organization. How can that be true?

Sovereignty over the world of science is vested in no particular ruler or governing body, but is divided into numerous fragments, each of which is wielded by one single scientist. Every time a scientist makes a decision in which he ultimately relies on his own conscience or personal beliefs, he shapes the substance of science or the order of scientific life as one of its sovereign rulers. The powers thus exercised may sharply affect the interests of his fellow scientists. Yet there is no need for a paramount supreme power to arbitrate in the last resort between all these individual decisions. There are divisions among scientists, sometimes sharp and passionate, but both contestants remain agreed that scientific opinion will ultimately decide right; and they are satisfied to appeal to it as their ultimate arbiter. Scientists recognize that, inasmuch as each scientist is following the ideals of science according to his own conscience, the resultant decisions of scientific opinion are rightful. This absolute submission leaves each free since each remains acting throughout in accordance to his own conviction. A common belief in the reality of scientific ideals and a sufficient confidence in their fellow scientists' sincerity thus resolves among scientists the apparent internal contradiction in the conception of freedom. It establishes government by scientific opinion, as a General Authority, inherently restricted to the guardianship of the premisses of freedom.

We are reminded of Rousseau's conception of liberty as absolute submission to the General Will. The devotion of all scientists to the ideals of scientific work may be regarded as the General Will governing the society of scientists. But this identification makes the General Will appear in a new light. It is seen to differ from any other will by the fact that it cannot vary its own purpose. Scientists who would suddenly all lose their passion for science and take up instead an interest in greyhounds would instantly cease to form a scientific society. The co-operative structure of scientific life could not serve the purpose of the joint breeding of greyhounds, for the pursuit of which the former scientists would have to organize themselves once more quite afresh. Scientific society is not and cannot be formed by a group of persons taking first the decision of binding themselves to a General Will and then choosing to direct their general will to the advancement of science. Scientific life illustrates on the contrary how the general acceptance of a *definite* set of principles brings forth a community governed by these principles—a community which would automatically dissolve the moment its constitutive principles were repudiated. The General Will appears then as a rather misleading fiction; the truth being (if the case of science be a guide) that voluntary submission to certain principles necessarily generates a communal life governed by these principles, and that ultimate sovereignty then rests safely with each generation of individuals who, in their devotion to these principles, conscientiously interpret and apply them to the issues of the period.

This also throws a new light on the nature of the Social Contract. In the case of the scientific community the contract consists of the gift of one's own person—not to a sovereign ruler as Hobbes thought, nor to an abstract General Will as Rousseau postulated—but to the service of a particular ideal. The love of science, the creative urge, the devotion to scientific standards—these are the conditions which commit the novice to the discipline of science. By apprenticing himself to an intellectual process based on a certain set of ultimates, the newcomer enlists as a member of the community holding these ultimates and his commitment to these necessarily involves the acceptance of the rules of conduct indispensable to their cultivation. Each new member undertakes to follow through life an obligation to a particular tradition to which his whole person gives assent.

Since a scientist requires special gifts, lack of these voids the contract. So does also lack of true animus, as in the fraudulent or unsound novice. I have described the disciplinary methods by which the scientific community strives to keep out bunglers, frauds, and cranks and pointed out the grave problems involved in distinguishing from these the great pioneers of revolutionary portent, who desire to enter on the Social Contract of science under modified conditions from the start. However, the difficulties which may arise in this connexion cannot affect the essential clarity of the contract by which the scientist becomes a member of his community. It consists in his dedication to the service of a particular spiritual reality.

We have seen how this dedication, pledging him to act according to his own conscience, represents an obligation to be free. Freedom of this kind, it would seem, must be described in the particular as freedom to act according to particular obligations. Just as a person cannot be obliged in general, so also he cannot be free in general, but only in respect to definite grounds of conscience.

II

Let us now step outside science into the wider context of society and examine the kind of freedom which is required in order to decide competently whether to accept or reject science as a whole.

Throughout modern history science has made an immense impression on the general public, and this was as strong as ever, if not strongest, in the earlier centuries of modern science when the practical value of science had been little thought of. It was the intellectual quality of science—particularly of Newtonian mechanics—which roused and convinced wide circles. Looking back on the past four centuries we see every department of thought gradually revolutionized under the influence of the discoveries of science. The medieval approach of Aristotle and Aquinas aiming at the discovery of a divine purpose in the phenomena of nature has been abandoned and theology forced to withdraw everything that it had taught of the material universe. While the occurrence of certain miracles, particularly of the Incarnation and Resurrection, is affirmed, Protestant theology is prepared to reinterpret miracles in general in a symbolic sense rather than oppose specifically the naturalistic views of science. Belief in witchcraft—still strong in the early eighteenth century—has been abandoned and astrology has been deprived of all official support. The current outlook on man and society has been transformed.

These conquests of science have been achieved at the expense of other mental satisfactions which proved the weaker. While the world has been enriched in one form of meaning it has inevitably lost some of its meaning in other forms. Galileo himself, though foremost in the attack against Aristotle's authority, showed real sympathy for the pain which he knew to be causing to those cherishing a belief in the great harmonies of scholasticism. No wonder then that the mental desires which science leaves unsatisfied have always been prepared to return to the charge. Thus for example Christian Science succeeds in contesting effectively even to-day the interpretation of disease and healing by science. A number of other unorthodox schools of healing flourish widely. Other theories condemned by science, such as those of astrology and occultism, are also upheld by a considerable public. The popular authority of science remains in fact open to challenge by various rival interpretations of nature, and the question remains how such rivalries can be competently decided.

A controversy between two fundamentally different views of the same region of experience can never be conducted as methodically as a discussion taking place within one organized branch of knowledge. While clashes between two conflicting scientific theories or two divergent biblical interpretations can usually be brought to a definite test in the eyes of their respective professional opinions, it may be extremely difficult to find any implications of a naturalistic view of man on the one hand and of a religious view on the other, in which these two can be specifically contrasted in identical terms. The less two propositions have fundamentally in common the more the argument between them will lose its discursive character and become an attempt at mutually converting each other from one set of grounds to another, in which the contestants will have to rely largely on the general impression of rationality and spiritual worth which they can make on one another. They will try to expose the general poverty of their opponent's position and to stimulate interest for their own richer perspectives; trusting that once an opponent has caught a glimpse of these, he cannot fail to sense a new mental satisfaction, which will attract him further and finally draw him over to its own grounds.

The process of choosing between positions based on different sets of premisses is thus more a matter of intuition and finally conscience, than is a decision between different interpretations based on the same or closely similar sets of premisses. It is a judgement of the kind involved in scientific discovery. Volition may play an important part in such judgements. We recall that an inflexible will is essential in scientific research if intimations of discovery are ever to reach the stage of maturity; and that very often it is right to persist in certain intuitive expectations, even though a series of facts are apparently at variance with it. Yet through all these struggles our volition must never finally determine our judgement which must remain ultimately guided by the quiet voice of conscience. Similarly, the mental crises which may lead to conversion from one set of premisses to another are often dominated by strong impulses of will-power. Conversion may come to us against our will (as when faithful communists were overcome by doubts and broke down almost overnight at the aspect of the Russian trials), or—see the example of St. Augustine—it may be vainly sought for years by the whole power of our volition. Whether our will-power be evoked by our conscience to assist its arguments or drive us on the contrary in a direction opposed both to argument and conscience, no honest belief can be made or destroyed—but only self-deception induced—by will-power alone. The ultimate decision remains with conscience.

This finally brings us up against the question: what premisses will guide conscience in decisions of this kind in a free society? Can we find, as in the case of the premisses of science, a practical art which embodies them; a tradition by which this art is transmitted; institutions in which it finds shelter and expression? Yes, we shall find them underlying the art of free discussion, transmitted by a tradition of civic liberties and embodied in the institutions of democracy. This art, this tradition, these institutions will be discovered in their purest forms in countries like Britain, America, Holland, Switzerland, where they were first and most effectively established.

I can see two main principles underlying the process of free discussion. One I will call fairness, the other tolerance, the words being used in a somewhat particular sense.

Fairness in discussion is the effort to put your case objectively. When an expression of our conviction first comes to our minds it is couched in question-begging terms. Emotion breaks out uppermost and permeates our whole idea. To be objective we must sort out facts, opinions, and emotions and present them separately, in this order. This makes it possible for each to be separately checked and criticized. It lays our whole position open to our opponent. It is a painful discipline which breaks our prophetic flood and reduces our claims to a minimum. But fairness requires this; and also that we ascribe our opponent his true points, while the limitations of our own knowledge and our natural bias be frankly acknowledged.

By tolerance I mean here the capacity to listen to an unfair and hostile statement by an opponent in order to discover his sound points as well as the reason for his errors. It is irritating to open our mind wide to a spate of specious argument on the off-chance of catching a grain of truth in it; which, when acknowledged, would strengthen our opponent's position and be even unfairly exploited by him against us. It requires great strength of tolerance to go through with this.

In the maintenance of fairness and tolerance the wider public plays a great part. Controversies between leaders of thought are usually conducted in order to canvass supporters rather than to convert each other. Fairness and tolerance can hardly be maintained in a public contest unless its audience appreciates candour and moderation and can resist false oratory. A judicious public with a quick ear for insincerity of argument is therefore an essential partner in the practice of free controversy. It will insist upon being presented with moderate claims admitting frankly their element of personal conviction. It will demand this both in order to defend the balance of its own mind and as a token of clear and conscientious thinking on the part of those canvassing its support.

The principal spheres of culture usually appeal as a whole

to the public, which as a rule accepts or rejects the opinion 'of science' or the teachings 'of religion' in their entirety without trying to discriminate between the views of different scientists or of different theologians. Yet occasionally they will intervene even in the internal question of one or the other great domain of the mind, particularly where an altogether new point of view is in rebellion against the ruling orthodoxy. Cultural rebels usually stand with one foot outside a recognized sphere, trying to get a hold in it with the other. Some parts of the public will come to their aid, others decry their efforts. The rise to scientific recognition in our own time of psycho-analysis, manipulative surgery, and most recently of telepathy, owe much to popular support. On the other hand, popular intervention, for example, of nationalist French circles demanding recognition for the Glozel finds, or of German anti-Semitic students opposing Einstein's theory of relativity, was wrong. Generally speaking, intervention by the general public when made in sincere search for the truth will be considered as rightful in a liberal society, provided it is kept within limits so as not to impair the sphere of autonomous government accorded to the experts under the protection of the community as a whole.

This brings us to the institutions which give shelter to free discussion in a free society. In Britain, for example, there are the Houses of Parliament; the courts of law; the Protestant churches; the press, theatre, and radio; the local governments, and the innumerable private committees governing all kinds of political, cultural, and humanitarian organizations. Being of a democratic character, these institutions are themselves guided by a free public opinion. Discussion is particularly protected for this purpose throughout their own body, rules of fairness and tolerance being enforced by custom and law. A wide range of divergent opinions is similarly protected throughout society at large. It is true that the status afforded to these varies greatly. Some, for example science, are given positive support both to develop further and to teach their doctrine widely. Other opinions, for example magic and astrology, are correspondingly discouraged.

Even though not all opinions are equally tolerated, protection is granted to many which cause pain and annoyance to people who disagree with them. The balance between opinions which are positively fostered and others which are only tolerated, and others again which are discouraged or even regarded as criminal, is constantly in flux. The necessities of war, for example, may cause the range of tolerance to be sharply narrowed. Public opinion is constantly making adjustments in these matters by custom and legislation.

However, neither these institutional rules and still less the general principles of fairness and tolerance, can be given the form of unequivocal prescriptions. Even the most stringently controlled field of discussion as formed by the procedure of the law courts leaves a margin for discretion. Borderline cases or fundamentally novel situations will frequently call for new interpretative judgements. In the wide fields of public argument each participant has to interpret day by day the existing custom in the light of his own conscience. These innumerable independent decisions would result in chaos but for the essential harmony prevailing between the individual consciences in the community. This consensus of consciences is usually described as showing the presence of a democratic spirit among the people. In the light of the previous analysis we can lay down more definite conditions for it.

In this light the 'democratic spirit' which guides the life of a free nation appears—like the scientific spirit underlying the activities of the scientific community—as an expression of certain metaphysical beliefs shared by the members of the community. They have been adumbrated already; we shall now turn to their analysis.

Fairness in discussion has been defined as an attempt at objectivity, i.e., preference for truth even at the expense of losing in force of argument. Nobody can practise this unless he believes that truth exists. One may, of course, believe in truth and yet be too biased to practise objectivity; indeed there are a hundred ways of falling short of objectivity while believing in truth. But there can be no way of aiming at the truth unless you believe in it. And furthermore there is no purpose in arguing with others unless you believe that they also believe in the truth and are seeking it. Only in the supposition that most people are disposed towards truth essentially as you are yourself is there any sense in opening yourself up to them in fairness and tolerance.

A community which effectively practises free discussion is therefore dedicated to the fourfold proposition (1) that there is such a thing as truth; (2) that all members love it; (3) that they feel obliged and (4) are in fact capable of pursuing it. Clearly these are large assumptions, the more so since they are of the kind which can be invalidated by the mere process of doubting them. If people begin to lose confidence in their fellow citizens' love of truth, they may well cease to feel obliged to pursue it at a cost to themselves. Considering how weak we all are at times in resisting temptation to untruthfulness and how imperfect our love of truth is at the best, it is the more surprising that there should exist communities in which mutual confidence in the sincerity of all should be upheld to the extent shown by their practice of objectivity and tolerance among themselves.

The love of truth and confidence in their fellows' truthfulness are not effectively embraced by people in the form of a theory. They hardly even form the articles of any professed faith, but are embodied mainly in the practice of an art—the art of free discussion—of which they form the premisses. This art—like that of scientific discovery which we studied before—is a communal art, practised according to a tradition which passes from generation to generation, receiving the stamp of each before being handed on to the next. There is a broad flow of this tradition which is passing through the whole of humanity but there are some more specific and elaborate forms of it which are carried on by single nations. The civic institutions of England have been the chief vehicles of this tradition since the seventeenth century. Dedication to the premisses of free thought means adherence to some national tradition in which similar institutions have taken deep root.

When a child is born to a national community the Social Contract is imposed on it by force. The community impels adherence in the first place by imparting a primary education in terms of its own premisses. A child growing up in a modern community will be forced to abandon the magical outlook to which it is primarily inclined and to adopt instead a naturalistic view of everyday life. In free communities it will be trained to practise fairness and tolerance. The whole heritage of free institutions will descend upon the youngster and confirm him in these traditional obligations. The premisses of freedom will thus be secured by compulsion, exercised by public opinion either directly or through the process of legislation.

It is hardly surprising that the Social Contract is so much less free for a nation than it is for the scientific community. There is plenty of scope outside science for those who have no love for it or else have to be removed from the scientific community for lack of ability or a breach of integrity. But a nation must absorb all those born in its midst, nor can it expel any of them later except by execution or exile. Moreover, members admitted to a community at birth cannot be given a free choice of their premisses; they have to be educated in some terms or other, without consultation of any preference of their own. In these circumstances the sense of obligation, by which the Social Contract is sealed, cannot but be firmly guided—if not induced altogether—by educational influence. We recognize herein the proper functions of the General Authority charged with upholding the premisses of free thought.

Nevertheless, every new member subscribing to a national (or general human) tradition adds his own shade of interpretation to it—and some will sign the contract only with far-reaching reservations. Each generation has the problem of sorting out the few great innovators from a multitude of cranks and frauds and has to decide this selection according to its own light. They must rely in the last resort on their own consciences. Whether a free nation endures, and in what form it survives, must ultimately rest with the outcome of individual decisions made in as much faith and insight as may be everyone's share. Any power authorized to overrule these decisions would of necessity destroy freedom. We must have sovereignty atomized among individuals who are severally rooted in a common ground of transcendent obligations; otherwise sovereignty cannot fail to be embodied in a secular power ruling absolutely over all individuals.

Atomized sovereignty, the sovereignty of a free public opinion, is also the resting-place on which the ultimate foundations of science are established. A community pledged to seek the truth cannot fail to accord freedom to science as one form of truth. Such adherence as it can gain by fair and tolerant public discussion is its rightful share. A scientist may ask for more: that is the part to be played by him in the contest of free competition; but as a citizen he will have to agree that such share as public competition establishes is rightful. This share may be determined to some extent by educational or other institutional action and still remain rightful, so long as such action is based on democratic decisions swayed by open persuasion.

This is the ultimate point to which we can trace the roots of our conviction expressed in affirming any particular scientific proposition as true. Such conviction implies in the last resort our adherence to a society dedicated to certain abiding grounds; among which are the reality of truth and our obligation and capacity to discover the truth. It affirms that in a society so dedicated a competent choice can be made between accepting or rejecting the premisses of science and that we have made that choice and accepted those premisses. And it goes on to affirm belief in the competence of the process of discovery which I described in the previous two lectures and in the validity, on the whole, of the results thus obtained. It finally sanctions one particular proposition by personally accrediting it in the light of all these premisses. By this last act is expressed also a belief that what is indicated by such a proposition is real; for which belief we also take personal responsibility. To this belief is linked the demand that the proposition be universally recognized as true. Thus, while we recognize that true propositions cannot be established by any explicit criteria, we do assert the universal validity of propositions to which we personally assent. Therein is expressed our conviction that truth is real and cannot fail to be recognized by all who sincerely seek it; and our belief in a free society as an organization of its members' consciences for the fulfilment of their inherent obligation to the truth.

Thus to accord validity to science—or to any other of the great domains of the mind—is to express a faith which can be upheld only within a community. We realize here the connexion between Science, Faith and Society adumbrated in these essays.

We may try to penetrate one step further by asking what the grounds are on which we hold the conviction that truth is real, that there is a general love of truth among men and a capacity to find it? These convictions (and others closely related to them, like the belief in justice and charity) have recently become involved in a fateful crisis. Our examination of the ultimate grounds on which our obligation to the truth rests will therefore quite naturally turn into an analysis of the general crisis in which our civilization is involved to-day.

This crisis has become most sharply manifest as a menace to all intellectual freedom based on the acceptance of a universal obligation to the truth. It would seem that it arose because the strictly limited nature of intellectual freedom had never been fully accepted by those who helped to establish it. They did not recognize that freedom cannot be conceived except in terms of particular obligations of conscience, the pursuit of which it permits and prescribes. They thought that freedom cannot mean the acceptance of any particular obligations and it is in fact incompatible with a prescription of its own limits. Freedom of thought in particular meant in their view the rejection of any kind of traditional beliefs, including, it would appear now, those on which freedom itself is based. They held that if any limits whatever were set to doubt, there would be no way of restraining intolerance and avoiding obscurantism.

Let me outline briefly the historical process by which our modern crisis has arisen.

III

The gradual emergence of a society dedicated to the pursuit of truth by the methods of objectivity and tolerance occurred in Europe through the revival of Greek thought after the Dark Ages. Much of this thought had survived in Christian theology and in the remnants of Roman Law. Then from the time of the Carolingian revival ancient thought spread its influence steadily until it once more became dominant during the Italian Renaissance. The period of the Renaissance humanists saw the first attempt to overthrow the hitherto ruling theological authority and to establish in its place a culture based on a free secular intelligence. The Reformation and the Counter-Reformation threw this process back, but it re-emerged finally in the seventeenth century in Holland, England, and in the English Colonies of America, and led there for the first time to an institutionally established régime of comparatively wide objectivity and tolerance. In other parts of Europe tolerance spread first through the agency of enlightened absolutism and later, more effectively, through the repercussions of the French Revolutions of 1789 and 1848.

The theological authority of the Medieval Church was severe and specific to a degree which seems intolerable to-day. As late as 1700 a good Catholic educated in France would be taught and would believe that our first ancestor Adam died on the 20th of August of the world year 930. All cases of doubtful interpretations of the faith were reserved to priestly authority. Compulsory annual confession backed up by the princes' sworn obligation to eradicate all heresy, as indicated to them by the Church, kept this régime firmly established throughout the latter Middle Ages.

The struggles which finally led to its general destruction have lasted up to our own times. They have produced our liberal forms of public life based on the assumption of the reality of truth and of the efficacy of reasoned argument. The medieval system founded on one specific text as interpreted by one central authority was replaced by a society founded on general principles interpreted by public opinion.

The new spirit of independence had been practised already for many years and in a variety of forms—artistic, political, religious, and scientific—before a resolute attempt was made to incorporate its premisses in a system of philosophy. Cartesian doubt and Locke's empiricism became then the two powerful levers of further liberation from established authority. These philosophies and those of their disciples had the purpose of demonstrating that truth could be established and a rich and satisfying doctrine of man and the universe built up on the foundations of critical reason alone. Self-evident propositions or the testimony of the senses, or else a combination of the two, would suffice. Both Descartes and Locke maintained their belief in the revealed Christian doctrine. And though the later rationalists succeeding them tended towards deism or atheism they remained firm in their conviction that the critical faculties of man unaided by any powers of belief could establish the truth of science and the canons of fairness, decency, and freedom. Thinkers like Wells and John Dewey, and the whole generation whose minds they reflect, still profess it to-day, and so do even those most extreme empiricists who profess the philosophy of logical positivism. They are all convinced that our main troubles still come from our having not altogether rid ourselves of all traditional beliefs and continue to set their hopes on further applications of the method of radical scepticism and empiricism.

It seems clear, however, that this method does not represent truly the process by which liberal intellectual life was in fact established. It is true that there was a time when the sheer destruction of authority did progressively release new discoveries in every field of inquiry. But none of these discoveries —not even those of science—were based on the experience of our senses aided only by self-evident propositions. Underlying the assent to science and the pursuit of discovery in science is the belief in scientific premisses to which the adherents and cultivators of science must unquestioningly assent. The method of disbelieving every proposition which cannot be verified by definitely prescribed operations would destroy all belief in natural science. And it would destroy, in fact, belief in truth and in the love of truth itself which is the condition of all free thought. The method leads to complete metaphysical nihilism and thus denies the basis for any universally significant manifestation of the human mind.

It might be objected that sceptics have in fact continued to love and uphold both science and its sister domains, as well as the régime of objectivity and tolerance in general. That is true—or at least quite frequently true. But it only shows that people can carry on a great tradition even while professing a philosophy which denies its premisses. For the adherents of a great tradition are largely unaware of their own premisses, which lie deeply embedded in the unconscious foundations of practice. These premisses can therefore remain long immune against their theoretical denial by those practising and transmitting the tradition. Thus science has been carried on successfully for the last 300 years by scientists who were assuming that they were practising the Baconian method, which in fact can yield no scientific results whatever. Far from realizing the internal contradiction in which they are involved, those practising a tradition in the light of a false theory feel convinced—as have been generations of empiricists descending from Locke—that their false theories are vindicated by the success of their right practice.

Such a state of suspended logic will, however, be less likely to develop in countries to which a tradition, not indigenous to its soil, is transmitted through its false theory rather than its true practice. This was shown to some extent already in France, where the unqualified conception of freedom derived from Locke's theories of government produced, quite logically, Rousseau's doctrine of absolute popular sovereignty: a doctrine which inaugurated Jacobinism and has hampered up to this day the practice of tolerant discussion between political parties in France. But even more serious were the consequences of a false doctrine of liberty percolating further east into countries with even less popular civic traditions. It became current there in the Romantic theories of the unrestrained individual and the unrestrained nation, and in the Socialist theory of the revolutionary class; all of which radically deny the possibility of objectivity and fairness in public discussion and give support, explicitly or by implication, to a totalitarian theory of the state. Nor did these theories remain on paper. While maxims of violence were advanced by writers on politics at all times, and since Machiavelli such precepts never ceased to affect the actions of statesmen, the twentieth century was the first in history to produce mass movements denying the reality of reason and equity and professing themselves to be actuated by sheer love of power.

These movements justified themselves by the support of supposedly scientific theories. This may appear illogical since they denied to science a position of independence; but it was true nevertheless. The class-war theory claimed that the rise of the working class to absolute power was scientifically inevitable. The Romantic theory affirmed it as a biological necessity that the superman and the super race shall achieve absolute mastery. Both Bolshevik and Fascist action were based on theories of unlimited violence; but the tribal and vitalistic element of Fascism led to a deliberate cult of brutality which was entirely absent from the purely mechanistic outlook of Bolshevism.

Both these movements, however, did not gain their great force from their professed sources of strength. We must not fall into their false view of man by accepting their own assessment of themselves. It was neither the acquisitive interests of the proletariat nor the physical vitality of the Italian and German peoples which carried the Bolshevik and Fascist revolutions to victory. These movements owed their success altogether to their hidden spiritual resources. They were swept into power on a tide of humanitarian or patriotic passions. The explanation seems clear enough. The denial of all spiritual reality is not only false but incapable of consummation. It is logically false to deny the existence of truth since the very statement asserting this is based on the assumption that truth can be established. But spiritual reality does not only continue to be implied in this sense but also remains an operative force. When we say 'truth is what benefits the proletariat' or 'truth is what benefits Germany' this does not cancel our conviction of truth or our love of truth, but merely transfers the transcendent obligations which we owe to truth to the temporal interests of the proletariat or the Germans. And the same holds for justice and charity for which our implicit attachment, like that for truth, is imperishable. Those who declare that these ideals have no real substance and that only the interests and power of particular groups are real, inevitably attach their aspirations for equity and brotherhood to the struggle of a particular party for power. Their ultimate reliance and all their love and devotion are then attached to this residue of reality, the power of the chosen party. Hence the selected party's irresistible fanaticism and its capacity to stir up deep moral response even while pouring scorn on moral realities.

From this analysis of its foundations we arrive at the following theory of totalitarian government. In order that a society may be properly constituted there must be competent forces in existence to decide with ultimate power every controversial issue between two citizens. But if the citizens are dedicated to certain transcendent obligations and particularly to such general ideals as truth, justice, charity, and these are embodied in the tradition of the community to which allegiance is maintained, a great many issues between citizens, and all to some extent, can be left—and are necessarily left—for the individual consciences to decide. The moment, however, a community ceases to be dedicated through its members to transcendent ideals, it can continue to exist undisrupted only by submission to a single centre of unlimited secular power. Nor can citizens who have radically abandoned belief in spiritual realities—on the obligations to which their conscience would have been entitled and in duty bound to take a stand—raise any valid objection to being totally directed by the state. In fact their love of truth and justice turn then automatically, as I have shown, into love of state power.

The dedication of a community to traditional ideals involves its assent to social action serving these ideals. To that extent the community is therefore deflected from its own tangible interests. Governments founded on the denial of spiritual reality can regard such deflection only as irresponsible drifting which they must counteract by appropriate intervention in every relevant detail. That is why totalitarian planning is logically necessary and must be comprehensive.

As applied, for example, to science, such planning means the attempt to replace the aims which science sets itself by aims set to science by the government in the interest of public welfare. It makes the government responsible for the ultimate acceptance or rejection by the public of any particular claims of science and for granting or withdrawing protection to particular scientific pursuits in accordance with social welfare. The proper aims of science being denied justification and even reality, the scientist still pursuing them is naturally held guilty of a selfish desire for his own amusement. It will be logical and proper for the politician to intervene in scientific matters, claiming to be the guardian of higher interests wrongly neglected by scientists. It will be sufficient for a crank to commend himself to a politician in order to increase considerably his chances of recognition as a scientist. In fields where scientific criteria allow wide latitude of judgement (e.g. medicine, agricultural science, or psychology) the crank who can enlist political support will find easy openings for establishing himself in a scientific position. Thus corruption or outright servitude will weaken and narrow down the true practice of science; will distort its rectitude and whittle down its freedom. And it will similarly distort and whittle down all rectitude and freedom in every field of cultural and political activity.

A society refusing to be dedicated to transcendent ideals

chooses to be subjected to servitude. Intolerance comes back full cycle. For sceptical empiricism which had once broken the fetters of medieval priestly authority, goes on now to destroy the authority of conscience.

IV

But I must not shut the gates of hope on the future. Totalitarianism has never been fully established in any place; as in fact no society could continue to exist for one day if the radical denial of spiritual reality were actually put into effect. Even though an organization had no other conscious purpose than to put sheer violence into operation and to exalt the supremacy of force over the spirit, it could never function without engaging for itself the support of idealist devotion. Besides, even though a community had at some time decided to live by a false idea of man, it might gradually forget this and be penetrated and finally absorbed once more by a renewal of cultural life and civic institutions stemming from its original civilization. For example in Soviet Russia, originally based on a class conception of society, we see pure science once more recognized, literature freed from Marxist interpretation, religion reinstated, national tradition revived, and the principles of private law gradually re-established. It is not inconceivable that a similar development might even have occurred in Nazi Germany a generation or two after Hitler's death.

But of course a very different line of future development may be approaching instead. The headlong descent of Europe from its peak of freedom and idealism achieved thirty years ago, down to its present state of conflict and violence, may presently gain new momentum by spreading to countries yet relatively untouched by it. Britain may not be able to uphold indefinitely the state of suspended logic which as yet protects her from the effect of the false theories current here as elsewhere. All these different eventualities rest ultimately with the consciences of men, for the enlightenment of which we may pray, but the decisions of which it is not for us to foresee.

Furthermore, I must make it clear that I have not intended to refute here the position of metaphysical nihilism by pointing out that its general acceptance logically implies a totalitarian form of society. A doctrine which denies reality to science and law, to the great arts, to religion, and to freedom in general, could well find the general destruction of these spiritual spheres acceptable in theory. Science, law, freedom, &c., could for example all be regarded as mere ideologies based on an outworn economic system, doomed to perish with this system. More savage doctrines than this have been taught at German universities and put into practice by their students.

Of course, believing as I do in the reality of truth, justice, and charity, I am opposed to a theory which denies it and I condemn a society which carries this denial into practice. But I do not assume that I can force my view on my opponents by argument. Though I accept truth as existing independently of my knowledge of it, and as accessible to all men, I admit my inability to compel anyone to see it. Though I believe that others love the truth as I do, I can see no way to force their assent to this view. I have described how our love of truth is usually affirmed by adherence to a traditional practice within a community dedicated to it. But I can give no reason why such a community, or its practice, should live—any more than why I should live myself. My adherence to the community, if given, is an act of ultimate conviction and remains so whether resulting from mature choice or mainly determined by early education. I can see a number of definite reasons for remaining attached, for example, to the tradition of pure science and of liberty of conscience, rather than to join an organization based on the principles of class war or Fascism. But again I know that my reasons cannot compel assent. Neither the Marxist's nor the Fascist's theory of man and society admits of common ground for argument between their adherents and the believer in transcendent reality.

Yet where the metaphysical believer cannot hope to convince, he may still strive to convert. Though powerless to argue with the nihilist he may yet succeed in conveying to him the intimation of a mental satisfaction which he is lacking; and this intimation may start in him a process of conversion. To the Marxist this would merely mean the withdrawal of his transcendent beliefs from their embodiment in a theory of political violence and their establishment once more in their own right. Such conversions have often happened in recent years. More difficult is the case of the Romantic nihilist whose cult of brutality tends to corrupt the very core of humanity in him. The combination of false teachings with a savage upbringing may make his conversion at the best very slow and uncertain. Yet I would still trust that the grounds for his conversion are there and expect to find in him a conscience which—once awakened—is as susceptible to its obligations as that of any man.

But I have yet to meet the objection that the position advocated here of holding beliefs which are admittedly not demonstrable could be used as a justification for a complete licence of beliefs, for arbitrariness, intolerance, and obscurantism. Men might say: 'If there is no demonstrable truth, I shall call true whatever I like, for example whatever is to my advantage to assert.' Or: 'If you admit that your belief of truth is ultimately based on your personal judgement, then I, the State, am entitled to replace your judgement by my own and determine what you shall believe to be true.' This, however, is not a correct reference to my position. Though I deny that truth is demonstrable, I assert that it is knowable, and I have said how. My position could be accused of leading to such general licence only if this condition could be shown to follow from a general assertion by everyone of the truth as he knows it in the light of his own conscience. But I cannot admit the possibility of such a result since the coherence of all men's consciences in the grounds of the same universal tradition is an integral part of my position. Those who are prepared to accept my conception of conscience and tradition will not fear any anarchy from a general acceptance of conscience as men's guide to the truth; while those who do *not* accept these meanings assume the position of the metaphysical nihilist which I have already discussed. This is as far as I can go in answering the question on what grounds my convictions of the reality of truth, and of our obligations to serve the truth, are held.

The views which I have put forward in these lectures differ in three important points from the universalism of the eighteenth century to which they try in general to revert. (1) I wholly accept the impossibility—finally demonstrated by logical positivism—of verifying any of the universal statements commonly held by men. This precipitates the crisis caused by sceptical empiricism and vastly extends its scope. (2) I do not assert that eternal truths are automatically upheld by men. We have learnt that they can be very effectively denied by modern man. Belief in them can therefore be upheld now only in the form of an explicit profession of faith. In my view this would be quite impracticable but for the existence of traditions which embody such professions and can be embraced by men. Hence tradition, which the rationalist age abhorred, I regard as the true and indispensable foundation for the ideals of that age. (3) I accept it moreover as inevitable that each of us must start his intellectual development by accepting uncritically a large number of traditional premisses of a particular kind; and that, however far we may advance thence by our own efforts, our progress will always remain restricted to a limited set of conclusions which is accessible from our original premisses. To this extent, I think, we are finally committed from the start; and I believe that this should make us feel responsible for cultivating to the best of our ability the particular strain of tradition to which we happen to be born.

In conclusion let me indicate a wider context to which my views seem to lead. I believe to have shown that the continued pursuit of a major intellectual process by men requires a state of social dedication and also that only in a dedicated society can men live an intellectually and morally acceptable life. This cannot fail to suggest that the whole purpose of society lies in enabling its members to pursue their transcendent obligations; particularly to truth, justice, and charity. Society is of course also an economic organization. But the social achievements of ancient Athens compared with those of, say, Stockport—which is of about the same size as Athens was—cannot be measured by the differences in the standard of living in the two places. The advancement of well-being therefore seems not to be the real purpose of society but rather a secondary task given to it as an opportunity to fulfil its true aims in the spiritual field.

Such an interpretation of society would seem to call for an extension in the direction towards God. If the intellectual and moral tasks of society rest in the last resort on the free consciences of every generation, and these are continually making essentially new additions to our spiritual heritage, we may well assume that they are in continuous communication with the same source which first gave men their society-forming knowledge of abiding things. How near that source is to God I shall not try to conjecture. But I would express my belief that modern man will eventually return to God through the clarification of his cultural and social purposes. Knowledge of reality and the acceptance of obligations which guide our consciences, once firmly realized, will reveal to us God in man and society.

APPENDIX

1. *Premisses of Science*

IN LECTURE II on 'Authority and Conscience' it will be explained that the premisses of science cannot be explicitly formulated, and can be found authentically manifested only in the practice of science, as maintained by the tradition of science. This should not be understood to deny the usefulness of analysing the premisses of science. Though no systematic attempt can be undertaken here to carry out such an analysis, I shall try at least to illustrate the kind of ultimate suppositions which scientists have relied on at different times. They will be seen to present remarkable diversity even though fundamentally based on common ground.

The conception of nature on which Copernicus relied for his speculations was borrowed from Pythagoras. It assumed the universe to be governed by numerical and geometrical rules, the divination of which was the task of science. Kepler's first planetary system (1596) stands out as an illustration of this approach. It was based on the fact, which was true within the range of Kepler's calculations, that the five regular solids (of identical edge) could be fitted between the spheres of the six then known planets so that each polyhedron was inscribed in the same sphere around which the next was circumscribed.[1] This system was renounced by Kepler in his later work in which he boldly abandoned the Pythagorean doctrine of circular orbits and uniform motion and expanded the mathematical view of nature inherited from Pythagoras to include all forms of mathematical functions. This approach was once more modified by Galileo as he transferred the study of mechanics from the skies to the earth. To Galileo we owe the assumption of a universe consisting of mass in motion, governed by the laws of mathematical dynamics. His programme was fulfilled and expanded by a great step when Newton included both Kepler's celestial and Galileo's terrestrial laws in one universal system of dynamics. From this achievement of Newton there originated the assumption—which was to predominate until the middle of

[1] C. Singer, *A Short History of Science*, p. 201.

the nineteenth century—that science might ultimately reduce all phenomena to the mechanics of some ultimate constituent particles. Thus Dalton started on his theory of chemical combinations from a particular aspect of these Newtonian presuppositions. 'It seems probable to me', Newton had written, 'that God at the beginning formed matter in solid, massy, hard, impenetrable, movable particles, of such sizes and figures and with such other properties, and in such proportion, as most conducive to the end to which he formed them. . . .' Dalton, who repeatedly quotes this passage, clearly counted the atomic structure of matter among the primary suppositions of science. Similarly, the two great laws of Conservation—those of Matter and of Energy—made their first appearance as axioms of a rational view of nature, both having been regarded, it would seem, as variants of the Newtonian outlook. The conservation of mass was propounded by Lavoisier with the statement that '. . . nothing is created in the operations either of art or of nature, and it can be taken as an axiom that in every operation an equal quantity of matter exists both before and after the operation . . . '.[1] While the Conservation of Energy was announced by Julius Robert Mayer '. . . as an axiomatic truth, that during vital processes a conversion only of matter as well as of force occurs, and that creation of either the one or the other never takes place'.[2]

The modern presuppositions of science which were to bear fruit in the great speculative triumphs of the twentieth century took shape gradually with the stepwise abandonment of feature after feature of this materialistic and mechanical picture. Faraday and Maxwell first strained this picture by adding to it the assumption of a ubiquitous 'field'. The electronic theory then shifted its ground further by demanding that electrical properties should be regarded as ultimate qualities, irreducible—in contrast to heat, sound, smell, etc.—to manifestations of mass in motion.

Other and even more important changes of the premisses of science were to follow. They seem to have been induced primarily by the philosophical critique of science originating from Ernst Mach.

[1] Quoted by Sherwood Taylor, *Science Past and Present*, p. 126.

[2] Quoted by Sherwood Taylor, l.c., pp. 244–5.

Mach's programme was to eliminate from scientific propositions all implications which were tautologous or otherwise thought to be essentially unverifiable. This purpose was carried on far beyond its original scope by Einstein's principle of relativity which axiomatically laid down the essential univerifiability of absolute motion and demanded a conceptual framework in which the question of absolute motion was logically excluded. From this conceptual reorganization there emerged an essentially new set of propositions which yielded a rich harvest of new valid predictions. A new 'epistemological' method of speculative discovery was thus established. This method was applied by Einstein both in his theory of special relativity (1905) and of general relativity (1916). It played a great part in Heisenberg's formulation of quantum mechanics (1925) which started from an attempt to eliminate all non-observable implications from the existing quantum theory of atomic processes due to Bohr. The lead given by Einstein's work on relativity has also determined—ever since Weyl's pioneer attempt of 1918—the continued search for a 'general field theory', which eventually became directed at a unitarian conception of the 'field' from which gravitational, electrical, and mesonic fields could all be derived as special consequences (Schrödinger, 1943). Another form of the same endeavour culminated in the efforts of Eddington and Milne to derive a system of natural laws purely from premisses of reason. The profound modification of the premisses of science involved in this line of inquiry became particularly clear through the controversy which it aroused. The early reaction of scientists to Eddington's views may be judged from the fact that his derivation of the 'fine structure constant' $hc/2\pi e^2 = 137$ was caricatured in a fictitious communication to *Naturwissenschaften* (1931) among whose authors we meet a young physicist who has since reached great distinction in science. Nor has antipathy to the premisses of Eddington abated up to this day. Quite recently, an eminent English mathematician, while telling me of some new increasingly accurate confirmations of Eddington's prediction of the mass ratio of proton and electron (as the ratio of the two roots of the quadratic equation

$$10x^2 - 136x + 1 = 0),$$

confessed himself rather worried about this fact, as he thought that Eddington's views were undermining the true empirical approach to nature.

A brief digression may be permitted here. The successes of the 'epistemological' method have much strengthened the authority of the positivist conception of science among scientists. This result represents, in my opinion, an error of judgement. The positivist movement was undoubtedly justified and successful in pressing for the purification of science from tautologies and unwarranted implications, but the great discoveries resulting from this process cannot be credited to any purely analytical operation. What happened was that scientific intuition made use of the positivist critique for reshaping its creative assumptions concerning the nature of things. Nor was science thereby effectively reduced to a set of definitely verifiable statements as postulated by the positivist conception of science; but was revealed on the contrary as possessing a faculty of speculative discovery which strikingly refutes that conception.

Parallel to the positivist movement there has occurred in our time yet another transformation of the premisses of science. Earlier conceptions of reality, capable of visual presentation in space, were replaced by purely mathematical concepts (like multi-dimensional wave functions) signifying certain probabilities and determining certain energies, but having no conceivable pictorial meaning attached to them. 'Nature's fundamental laws', wrote Dirac in 1935, 'do not govern the world as it appears in our mental picture in any very direct way, but instead they control a substratum of which we cannot form a mental picture without introducing irrelevancies.' That substratum can be described only in mathematical terms. This feature of modern science had made its first appearance in Planck's quantum theory of 1900. It reappeared in all the various applications of quantum theory but was not definitely accepted as a basic element of science until, about 1925, it was organically embodied in the new quantum mechanics.

These illustrations may suffice to show how a number of marked variations took place during the past 400 years in the fundamental guesses of science concerning the nature of the universe. The picture of these changes which we have given is far from complete, for even though physics may justly be regarded as the most fundamental part of natural science it does not in fact form the operative premiss of either chemistry or biology. These are founded on their own basic suppositions which have also undergone a gradual historical development. There is in fact no aspect of science, including even mathematics, in which the fundamental presuppositions, the methods of investigation, and the criteria used for verification have not undergone a series of marked changes since the inception of modern science 300 years ago.

It is common enough therefore to come across statements by great scientists of the past which are quite unacceptable to modern scientists. Many of the arguments of Copernicus, Galileo, Kepler, Newton, Lavoisier, Dalton, seem irrelevant to-day and often we find that their presuppositions have led them to conclusions which we now consider to be false.

It is frequently said that the facts of science remain and only the interpretations change. This is not true or is at least very misleading. If we still recognize many of the facts which were collected, say by astronomers, 300 years ago, it is because in these cases we share their basic interpretation of the sensory experience which they described as facts. But while to Kepler in 1596 it appeared as an indubitable fact that the planetary orbits are related to the geometry of perfect solids, we regard this to-day as mere fancy. Or to take another example: Newton observed that even after repeated distillation water always left a slight residue behind and described it as a fact that water on evaporation is partly transmuted into earth. Though we accept Newton's experience as true, and could reproduce it in similar circumstances, we do not now consider that it established the fact which he claimed to have observed. Apart from meaningless sense impressions there is no experience that abides as a 'fact' without an element of valid interpretation having been imparted to it. This is true even of facts of everyday life, the nature of which depends on the accepted interpretation of events—whether magical, astrological, mythological, naturalistic, etc.

We may take it therefore that, in view of the changed premisses of science, much of earlier science appears to-day both factually and theoretically false. But it is even more obvious that much of earlier science is accepted to-day as true.

In fact the great pioneers of science keep growing in our respect through the centuries, as the significance of their discoveries becomes ever more broadly manifest. There must be therefore considerable common ground between the modern scientist and his earlier forerunners. In other words, the modern premisses of science include a great deal of the earlier premisses; enough at any rate to make us find many important conclusions which were originally drawn from those premisses wholly acceptable to us to-day.

In view of the nature of science, as described in these present lectures, we hold that no exhaustive statement of the premisses of science can possibly exist. The common ground of science is, however, accessible to all scientists and is accepted by them as they become apprenticed to the traditional practice of science. This thesis will be found elaborated particularly in my second and third lecture.

2. *Significance of New Observations*

The scientist in pursuit of research has incessantly to make decisions whether to take a new instrument reading or some other new sense impression as signifying a new fact, or to regard it merely as a new indication of an old fact—or else to reject it as having no significance at all. These decisions are guided by the premisses of science and more particularly by the current surmises of the time, but ultimately there always enters an element of personal judgement.

Some examples may illustrate these relationships.

It has been long accepted as a law of nature that—apart from the ascertained planets—all stars retain their positions to one another from one day to the next. Actually stars are never observed to be exactly in the same position one day as they had been the day before, but this is usually allowed for by the assumption of observational error. Consequently when a new planet is first observed, its motions will tend to be explained away as observational errors. When Neptune was discovered in 1846 the past positions of this planet were computed and its identity was established with a star recorded by Lalande in Paris in May 1795. This being communicated to the Paris observatory, an examination of Lalande's manuscript showed that he had made two observations of the planet, on the 8th and 10th of May and finding them discordant had rejected one as probably in error, and marked the other as questionable.[1] The planet Uranus before its actual discovery by Sir William Herschel in 1781 had been recorded as a fixed star at least seventeen times. Thus the routine process of reaffirming the known laws of nature becomes the mass grave of many a potential discovery.

Observations which can be interpreted as a transmutation of chemical elements frequently occur in the laboratory. But actual claims by reputable investigators of having achieved transmutation appear only at times when the possibility of such a process is for some reason considered plausible. In earlier times when the assumptions of alchemy were generally accepted by scientists, such claims were of course quite common. Newton considered the fact that water, even after repeated distillation, still left behind on evaporation a slight earthy residue as a proof for the spontaneous transformation of part of the water into earth. Observations of a similar kind no doubt continued to be made throughout the centuries, but since the acceptance at the end of the eighteenth century of Lavoisier's views on the nature of the elements they were explained as mere dirt-effects. Such at least was the case up to the beginning of the twentieth century. Then, suddenly, under the stimulus of Rutherford's and Soddy's discovery of radioactive transmutation (1902–3), a series of erroneous claims were made by careful observers to have achieved in their own way a transmutation of elements. A. T. Cameron (1907) and Sir William Ramsay (1908) announced the transformation of copper into lithium as a result of the action of α-particles. In 1913 Collie and Patterson claimed the formation of helium and neon by electric discharge through hydrogen. After these claims were disproved, no new ones appeared till 1922, when the discovery made three years earlier by Rutherford of certain forms of artificial transmutation encouraged a new wave of similar claims based on erroneous evidence. The transmutation of mercury into gold under the effect of electric discharges was reported quite independently by Miethe and Stammreich in Germany and Nagaoka in Japan. Smits and Karssen reported the transformation of lead into mercury and thallium. Paneth and Peters claimed the transformation of hydrogen into helium

[1] T. E. R. Phillips, *Enc. Brit.*, 14th ed., vol. xvi, p. 228.

under the influence of a platinum catalyst. All these claims, however, had to be abandoned in the end. The last of them was given up in 1928. A year later came the establishment of the theory of radioactive disintegration which showed clearly that the attempts described above to transform elements had been futile. Since then, up to this date no new claims were made in this direction although evidence of transformation of the kind put forward by Newton, Ramsay, Paneth, etc., is always at hand. It is now disregarded because it is no longer considered as sufficiently plausible.

Naturally, this is not to say that scientists are bound always to explain away observed deviations from hitherto accepted assumptions—which would make all scientific progress impossible. They may brush aside discrepancies as mere freaks or on the contrary attribute the greatest significance to them. Rutherford's genius has been well characterized in this connexion by one who knew him closely.[1] He could throw aside as irrelevant a stream of reports pouring in from all over the world about new oddities to which fellow scientists called his attention, and yet respond to one particular instance among them, raising a hue and cry such as caused Chadwick to discover the neutron. The well-known stories of Bequerel discovering radioactivity and Röntgen discovering X-rays by pursuing the clue of accidentally fogged photographic plates—which earlier observers had disregarded—also illustrate this kind of ability. We shall appreciate the courage and vision shown by such discoverers even better by bearing in mind the less well known but actually much more numerous cases when their mode of action led to failure—the lives wasted for example on investigating the spurious 'N-rays' (1902–6) and other such fictitious phenomena[2] which were stimulated by the very examples of Bequerel and Röntgen.

The problem of attributing to observations the right significance in respect to an existing framework of theory extends, of course, far beyond the decision of what to put down or not to experimental error. Certain observations may be recognized as establishing formal contradictions to a theory and yet be set aside for the time being. The two examples mentioned in the

[1] C. G. Darwin, *Nature*, vol. cxlv, p. 324.

[2] Comp. G. F. Stradling, *Journ. of the Franklin Inst.*, vol. clxiv, pp. 57, 113, 177.

text—the contradiction to the periodic system and the conflict between wave theory and quantum theory of light—are both instances in which the conflict was subsequently eliminated by the discovery of a more fundamental approach which accounted for both sides of the evidence. But this must obviously not be assumed to hold generally. Theories have quite often been abandoned on account of contradictory evidence and have vanished without leaving a trace behind. There is no need to give examples for that. But it may be of interest to recall that theories have sometimes been abandoned on account of contradictory observations and yet were later revindicated by further discoveries. The approximately integral atomic weights of the lighter elements helium, carbon, nitrogen, oxygen, etc., based on hydrogen, had convinced Prout that all elements were built of hydrogen. But the subsequent exploration of atomic weights, particularly of the heavier elements, convinced later scientific opinion that the deviations from integrality were too great and too numerous to allow Prout's hypothesis to be upheld. This decision must to-day be regarded as erroneous, as Prout's hypothesis has since been saved by the addition to it of the theory of isotopes and of packing effects.

The elimination of formal contradictions to a theory need not require the help of new discoveries. Some theories are built in such a way that the necessary amplifications can be introduced automatically; as when the motion of planets was described by cycles and epicycles and any deviation could be accounted for by introducing further elements of the kind. This is tantamount to the addition of a further term to a mathematical series by which certain observations are to be represented. Theories which are thus self-amplifying are sometimes called epicyclical (e.g., in genetics). This by no means disqualifies them as expressions of natural laws. It is true that they cannot be formally contradicted by any conceivable observation and can therefore strictly speaking predict nothing. But we have seen that this is true of all scientific propositions. All theories are 'epicyclical' in the sense that reasons are always conceivable which will account for an observed deviation. It always remains for the scientist to decide in the light of the general premisses of science, and of the particular assumptions considered plausible at the time, what weight to attach to any given set of observations in support or refutation of a theory on which they seem to be bearing. Ultimately this is a matter for his personal judgement.

3. *Correspondence with Observation*

The following example may illustrate that occasionally the most rigorous criteria of experimental verification may be fulfilled, and yet eventually prove to be fictitious products of a strange coincidence.

One of the most beautiful apparent confirmations of a scientific theory was the measurement by Aston in the mass spectrograph of the atomic weight of hydrogen and oxygen which (for oxygen assumed at 16.00000) gave H = 1.00778, as compared with the result obtained by chemical analysis H = 1.00777.[1] The correspondence was apparently made safe beyond any reasonable doubt when Bainbridge confirmed Aston's value (so far as this was based on the ratio He/H) by finding He/H = 3.97128 as compared with Aston's He/H = 3.97126. Bainbridge used a spectroscopic method, which is entirely different in its assumptions from that of Aston. This threefold set of accurate correspondence may have seemed unassailable; and yet its accuracy turned out to be quite accidental. First it was discovered that oxygen contained a slight admixture of heavier isotopes (O^{17} and O^{18}). Taking this into account, the chemical evidence now led to the expectation of a ratio O/H = 1.00750 in the mass spectrograph, and the accuracy of the previously observed correspondence (1.00777 and 1.00778) was destroyed. The new discrepancy led to the assumption that hydrogen also contained some heavier isotopes—and starting off on this clue Urey made a search for heavy hydrogen and discovered its presence in minute traces (1932). Urey's discovery was described at the time as a triumph of faith; which remains true, even though the faith to which it so bravely entrusted itself and which it so brilliantly vindicated proved false. Three years later Aston revised his earlier measurements and gave O/H = 1.0081. Such a value would, after the discovery of the heavy oxygen isotopes, correspond to a chemical atomic weight ratio of 1.0078 which does not require for its explanation the presence

[1] F. W. Aston, *Proc. Roy. Soc.*, A. cxv. 487 (1927).

of heavy hydrogen but would rather suggest that no such admixture was present.

Apart from such accidental coincidences which may lead to apparent confirmation of a false proposition in science, we must remember that our reliance on reproducibility suffers from a fundamental weakness. It is always conceivable that reproducibility depends on the presence of an unknown and uncontrollable factor which comes and goes in periods of months or years and may vary from one place to another. Take the following examples. In 1922 I observed jointly with H. Mark that when tin crystals, in the form of wires, were strained there appeared on their surface a set of characteristic slip lines.[1] Hundreds of such specimens were produced and some of them photographed and their pictures published. Identical photographs were published by C. Burger[2] who had independently made the same discovery. The investigation was carried on for a number of years in my laboratory, but after about 1923 no slip bands were ever observed; the crystals showed the same mechanism of slip, but their surface remained completely smooth. There has never been found any explanation for this changed behaviour. One is reminded—to take for once an example from the field of biology—of the mysterious loss of smell of the musk plant which seems to have occurred a few years ago suddenly all over the planet.

There are a large number of phenomena, such as the explosion of gases, the strain and breaking of solids, the electric breakdown of dielectrics, surface catalysis, crystallization, and electro-deposition, which depend on the trigger action of small traces or flaws. We know also that even the purest substances available in our laboratories contain traces, of say one part in a thousand million, of practically every chemical element. Any phenomena, therefore, which depend on the presence of certain substances in traces, may reproducibly maintain a certain character for a time and then suddenly take on—once more reproducibly for another period—a different one, depending on periodic variations of the elements present in traces. Instances of this are well known as 'epidemics' which affect the course of

[1] Mark and Polanyi, *Z. Phys.*, xviii. 75 (1923).

[2] Comp. C. Burger, *Physica*, i. 214 (1921); ii. 56 (1922).

industrial processes. They may come and go without the cause having been discovered.

Mr. R. G. R. Bacon of the Research Laboratories of I.C.I. Dyestuffs Division, Blackley, has recently given me some details of the following experience of his own illustrating this point:

About two years ago I was making numerous measurements of the rate of polymerisation of a vinyl-type monomer under nitrogen in an aqueous solution containing the reduction activation system persulphate/bisulphite as catalyst. Under a standard set of experimental conditions I observed a fast rate of polymerisation, which I had no difficulty in reproducing whenever I carried out the reaction. I left this work, but about a year later it was re-opened by another worker, who repeated my experiments but observed a very much slower rate of polymerisation. We therefore collaborated for a short time to see where the discrepancy lay. It was found that:

(*a*) I could reproduce my earlier results even when using the same reagents as the second chemist.

(*b*) My results were still reproducible when I carried out polymerisation in a new apparatus, similar in design and dimensions to that of the second chemist.

(*c*) When we worked simultaneously, drawing each one of our reagent solutions from a common supply, the second chemist still observed a very much slower polymerisation rate than I did.

(*d*) The differences in effect were not apparently due to differences in oxygen content, since the reaction was relatively insensitive to oxygen, and proceeded at much the same rate when the nitrogen atmosphere was replaced by air.

(*e*) The second chemist observed the same high rate of polymerisation as I did when, instead of using rubber tubing to carry his nitrogen supply, he used glass; in my own apparatus the nitrogen supply came through a metal tube (made of a soft lead alloy).

At this point the second chemist left our Department and we never carried out any further experiments to verify that the cause of the discrepancy really lay in his use of rubber tubing. I may mention that both before and after his experiments I had reproduced my high rate of polymerisation even when my own apparatus contained a rubber nitrogen lead, so the apparent effect of rubber was not a general one.

Experiences of this kind should remind us that there is always a *conceivable* doubt as to the convincing power of reproducibility; it is for the scientist to decide in the light of his own judgement whether to consider such doubt as *reasonable* in any particular instance.